Springer Theses

Recognizing Outstanding Ph.D. Research

For further volumes:
http://www.springer.com/series/8790

Aims and Scope

The series "Springer Theses" brings together a selection of the very best Ph.D. theses from around the world and across the physical sciences. Nominated and endorsed by two recognized specialists, each published volume has been selected for its scientific excellence and the high impact of its contents for the pertinent field of research. For greater accessibility to non-specialists, the published versions include an extended introduction, as well as a foreword by the student's supervisor explaining the special relevance of the work for the field. As a whole, the series will provide a valuable resource both for newcomers to the research fields described, and for other scientists seeking detailed background information on special questions. Finally, it provides an accredited documentation of the valuable contributions made by today's younger generation of scientists.

Theses are accepted into the series by invited nomination only and must fulfill all of the following criteria

- They must be written in good English.
- The topic of should fall within the confines of Chemistry, Physics and related interdisciplinary fields such as Materials, Nanoscience, Chemical Engineering, Complex Systems and Biophysics.
- The work reported in the thesis must represent a significant scientific advance.
- If the thesis includes previously published material, permission to reproduce this must be gained from the respective copyright holder.
- They must have been examined and passed during the 12 months prior to nomination.
- Each thesis should include a foreword by the supervisor outlining the significance of its content.
- The theses should have a clearly defined structure including an introduction accessible to scientists not expert in that particular field.

Pegor Aynajian

Electron–Phonon Interaction in Conventional and Unconventional Superconductors

Doctoral Thesis accepted by
Max Planck Institute for Solid State Research,
Stuttgart, Germany

Springer

Author
Dr. Pegor Aynajian
Department of Physics
Princeton University
Princeton 08544
USA
e-mail: aynajian@princeton.edu

Supervisor
Prof. Bernhard Keimer
Max Planck Institute for Solid State Research
Heisenbergstr. 1
70569 Stuttgart
Germany
e-mail: b.keimer@fkf.mpg.de

ISSN 2190-5053 e-ISSN 2190-5061

ISBN 978-3-642-14967-2 e-ISBN 978-3-642-14968-9

DOI 10.1007/978-3-642-14968-9

Springer Heidelberg Dordrecht London New York

Library of Congress Control Number: 2010935604

Cover design: eStudio Calamar, Berlin/Figueres

Printed on acid-free paper

Springer is part of Springer Science+Business Media (www.springer.com)

To My Parents

Supervisor's Foreword

The problem of conventional, low-temperature superconductivity has been regarded as solved since the seminal work of Bardeen, Cooper, and Schrieffer (BCS) more than 50 years ago. However, the theory does not allow accurate predictions of some of the most fundamental properties of a superconductor, including the transition temperature and the energy gap on the Fermi surface, because they depend exponentially on materials-specific parameters that cannot be computed with sufficient precision. This thesis describes a new experimental method that puts this old problem into an entirely new light. The method is based on ideas going back more than 30 years, but technical difficulties had foiled attempts to implement it. Under the leadership of Dr. Thomas Keller, our group has undertaken a major, and ultimately successful effort to overcome these obstacles, to which the author of this thesis has made major contributions. In rough analogy to nuclear magnetic resonance techniques, the new method takes advantage of the "spin echo" of neutrons scattered by a superconducting crystal in radio-frequency magnetic fields. In this way, the energy resolution of spectroscopic experiments on dispersive lattice-vibrational excitations (the "glue" responsible for Cooper pairing of electrons in conventional superconductors) can be enhanced by more than two orders of magnitude. Using this method,the author has discovered an unexpected relationship between the superconducting energy gap and the geometry of the Fermi surface in the normal state, both of which leave subtle imprints in the lattice vibrations that could not be resolved by conventional neutron spectroscopy (P. Aynajian et al., Science 319, 1509 (2008)). He has confirmed this relationship in experiments on two elemental superconductors and on a series of metallic alloys. This indicates that a mechanism qualitatively beyond the standard BCS theory determines the magnitude and an isotropy of the superconducting gap, at least in superconductors with sufficiently strong electron-lattice coupling. As a Ph.D. student, the author has initiated several experiments on binary (NbSe2) and quarternary ($La_{2-x}Sr_xCuO_4$) superconductors designed to test the generality of the relationship between the Fermi surface geometry and the energy gap, and the thesis describes preliminary results. These and other experiments are still ongoing, as are attempts to find a microscopic explanation of the observed effect. If these studies confirm the relationship as a general property of strong-coupling superconductors, a purely geometrical construction based on the Fermi surface and the dispersion relation of the lattice vibrations (both of which can be readily computed) can predict the magnitude and anisotropy of the superconducting energy gap. This would imply a major change of perspective on the old problem of conventional superconductivity.

Stuttgart, October 2010 B. Keimer

Acknowledgement

I would like to express my sincere gratitude to the numerous people whose contribution and support realized this thesis.

First of all, I am grateful for Bernhard Keimer who welcomed me to the Max Planck Institute and gave me the opportunity to work on the TRISP instrument. The time he spent with me for discussions are invaluable, which led to a profound understanding of many aspects of physics. I'm highly indebted for his support and interest in my work throughout the course of my stay, which highly motivated me. I'm also thankful for giving me the opportunity to present my research at the scientific advisory board meeting. Finally, I'm grateful for reading and correcting my thesis within a very short period, despite of all his obligations.

I'm also thankful for Harald Giessen, who kindly accepted to be the second reviewer of my thesis.

I'm very grateful for my direct supervisor, Thomas Keller, who handled me the TRISP instrument from its first operating days. That gave me confidence as an experimentalist to try different things regardless of the "real" experiment. I'm grateful for the time he took and explained the details of neutron spin echo spectroscopy beyond textbooks, which helped me in understanding and analyzing the data. I'm also grateful for all the discussions and his interesting ideas during the experiments. Even though he tried hard to teach me the German language so that I can write the summary of my thesis in German, in the end he was the one who translated the abstract.

I'm thankful for the scientific director of the FRM II, Winfried Petry, for his frequent visits during the experiments, excited to see "the phonon of the day". His support was very motivating.

Without the high quality samples that we got from Chengtian Lin, K. Yamada, Helmut Berger, Janos Mayor, and Andreas Erb, none of this work would have been realized.

I'm grateful for the collaboration I had with Lilia Boeri, who did all the ab initio calculations presented in this thesis. I would as well acknowledge the helpful discussions with Roland Zeiher, Oleg Dolgov, Giniyat Khaliullin, Oleg Sushkov,

Manuel Cardona, Klaus Habicht, Henrick Ronnow, Stefan Shapiroand, Klaudia Hradil and Ole Anderson which helped to interpret my results.

I'm thankful for Benjamin Bruha, Sibel Bayrakci, and Michael Schulz for the delicate high precision goniometer towers, and for Manfred Ohl and Heinrich Klann who helped me in all the technical stuff, and for our secretary Claudia Hagemann for her help in the several administrative procedures through out my stay in the Max Planck Institute.

Special thanks to Kathrin Buchner, who besides being a wonderful friend was very helpful during all the experiments, also for my colleagues and friends in the Max Planck Institute, Leininger Philippe, Bohnenbuck Britta, Nathalie Munnikes, Raichle Markus, Hinkov Vladimir, Yordanov Petar, Ulrich Clemens, Zegkinoglou Ioannis, Damljanovic Vladimir, Jackeli George, Haug Daniel, Matiks Yulia, Rahlenbeck Martin, Suchaneck Anton, Salman Majdi, Driza Nadir, Uhlig Heiko, Boris Alexander, Kovaleva Natalia, Inosov Dmytro, Popovich Paul, Benckiser Eva, Park Ji-tae and over all, Mohammed Bakr, with whom I spent most of my time at the Max Planck Institute and with whom I shared my coffee and cigarette every day.

I can't forget my lovely friends and at mates in Stuttgart, Fadi el Hallak, Jules Michael, Pedro Vidal, and Erhan Arac with whom I lived and shared my happiness and sadness. Without their constant support, motivation, and wisdom, I would have been far from realizing this work. Also my friends abroad, Eddy Toramanian, Hmayag Partamian, Sako Mikaelian, Apig Festekjian, Shiraz Toramanian, Talin Boyajian and many others, who believed in me and supported me from the first day I left my home country.

Finally I will thank my beloved wife Tamar, my parents, and my sister and brother, who enriched my life and encouraged me every single morning, and supported me in every possible way they could. Thank you.

Contents

Chapter 1
Introduction

1.1 History of Superconductivity

Superconductivity is a fascinating state of matter where electrons conduct a current without dissipation and where external magnetic fields are expelled from the sample. The history of superconductivity starts in 1911 when H. Kamerlingh Onnes discovered that the resistivity of Hg suddenly drops to zero below $T_c = 4.2$ K. Two decades later, Meissner and Ochsenfeld discovered that superconductors exhibit perfect diamagnetism, which today is considered as the most unambiguous signature of superconductivity [1]. Despite of intense experimental and theoretical research, a microscopic theory for superconductivity was founded only in 1957, almost five decades after its discovery, by Bardeen, Cooper, Schrieffer. [2], and is now known as the BCS theory. Superconductivity was explained by an effective attraction between electrons, mediated by lattice vibrations (phonons). These same phonons which scatter electrons at elevated temperatures leading to the electrical resistivity in metals, become the glue for electron-pairing at low enough temperatures, and lead to the formation of the so-called Cooper pairs [3]. Electrons are fermions, however in pairs they have a bosonic nature which allows them to condense into a single quantum state below the Fermi energy, thus opening a superconducting gap Δ on the Fermi surface. The strength of this pairing, the so-called electron–phonon interaction, defines the strength of the superconducting gap and therefore the transition temperature T_c of a superconductor.

Before the mid-1980s, superconductivity had only been observed in metals and metallic alloys with the highest observed transition temperature of 23 K. Due to the relatively low transition temperatures, technological applications of superconductors were severely limited. The breakthrough in the field of superconductivity however came in 1986, when Georg Bednorz and Alex Müller discovered that lanthanum copper oxide (La_2CuO_4), which is an insulator, when doped with barium, becomes a superconductor with a transition temperature, T_c, of 36 K [4]. Other materials, containing layers of copper and oxygen atoms alternated with layers of

P. Aynajian, *Electron–Phonon Interaction in Conventional and Unconventional Superconductors,* Springer Theses, DOI: 10.1007/978-3-642-14968-9_1,

other metals, with transition temperatures significantly exceeding the boiling temperature of nitrogen were discovered soon thereafter. Unlike the conventional superconductors, however, high-temperature superconductors could not be explained by the BCS theory in its original form. Even though the physical mechanism underlying pairing in high temperature superconductors (HTSC) is far from being understood, both HTSC and BCS superconductors involve Cooper pairs, which is a common property of superconductors independent of their pairing mechanism. Unlike BCS superconductors, where the superconducting gap has an s-wave symmetry (isotropic in momentum space), early experiments on HTSC indicated singlet Cooper pairing with a d-wave symmetry where the gap has nodes on the Fermi surface [5]. The earliest suggestions of a d-wave type pairing symmetry were based on the assumption that the pairing was achieved by exchange of antiferromagnetic spin excitations. Because of the appearance of the gap nodes, it was believed that the electron–phonon mediated pairing, which is supposed to be isotropic, could be ruled out. Recent theoretical considerations, however, indicate that this may not be true [6]. There has also been a recent accumulation of experimental results suggesting that the electron–phonon interaction does play an important role in these materials. Angle-resolved photoemission spectroscopy (ARPES) [7, 8], neutron scattering [9], scanning tunneling spectroscopy [10], and theoretical work [11, 12] have uncovered evidence of strong electron–phonon interactions in the cuprates. While much of this work is still controversial, it is undisputed that the electron–phonon interaction in materials with strong electronic correlations, such as the cuprates, is an interesting subject whose understanding is still in its infancy.

Recently, a new family of iron pnictide high-temperature superconductors with transition temperatures up to 52 K were discovered [13]. These materials resemble the cuprates in some striking ways. They are also layered materials, but instead of copper and oxygen, they contain planes of iron and arsenic. As in the cuprates and other unconventional superconductors, the parent compound is an antiferromagnet at low doping and increased doping destroys the antiferromagnetism leading to superconductivity. On the other hand, the conspicuous lack of the CuO_2 planes in the pnictides raises the question of a different pairing mechanism. Multiple gaps on the different Fermi surface sheets have been observed in experiments [13, 14] similar to that observed in MgB_2 [15]. The gaps decrease in temperature and close simultaneously at T_c consistent with the BCS prediction, but dramatically different from that of the pseudogap behavior in the cuprates. The nodeless and nearly isotropic gaps around their respective Fermi surfaces raise the possibility of a pairing mechanism different from that in cuprates, perhaps involving strong electron–phonon interaction as seen in MgB_2 [16].

The symmetry and the momentum dependent anisotropy of the superconducting gap is the key source of information on the microscopic mechanism of superconductivity. The anisotropy of the superconducting energy gap, Δ_k, of even the simplest elemental superconductors has been of long-standing interest not only as a testing ground for electronic structure calculations [17–21], but also as a possible source of information on electronic ordering phenomena that compete or coexist with superconductivity [22, 23]. A variety of experiments over a period of almost five decades have therefore targeted this quantity. However, the comparison of experimental and theoretical results, up to date, has been severely limited by experimental difficulties.

1.2 Neutron Spectroscopy of Superconductors

It was pointed out by Phillip Allen some time ago that the phonon linewidth is directly proportional to the electron–phonon coupling constant λ, and contains all the information about the electron–phonon interaction in superconductors [24]. For the past half a century, triple-axis spectrometry (TAS) with neutrons has been the method of choice to experimentally determine energy- and momentum-resolved phonon spectra of solids [25]. The energies and momenta of the incoming and scattered neutrons are selected by crystal monochromators, and their difference yields the phonon dispersion relation. The monochromaticity of the beam is thus coupled to the beam divergence, which can be restricted only at the expense of beam intensity. This implies that energy resolutions significantly better than 10% are impractical to achieve under almost all circumstances. Since the energy resolution required to detect the normal-state electron–phonon linewidth is typically a few μeV, phonon lifetime measurements were successful only in few exceptional cases.

The investigations reported in this thesis were made possible by the development of neutron resonant spin-echo (NRSE) spectroscopy, which enables the determination of the lifetimes of dispersive excitations with μeV energy resolution over the entire Brillouin zone [26–28]. The spin echo is implemented on a triple-axis spectrometer by using radio-frequency magnetic fields inserted between the monochromator and sample and between the sample and the analyzer. The extremely high resolutions are obtained by monitoring the change in the neutron spin of an initially polarised beam before and after the scattering process. This new generation spectroscopy allowed us to study the electron–phonon linewidths in conventional and unconventional superconductors extended over large parts of the momentum space and with an energy resolution in the μeV range.

Asking most scientists if it is worth to have a closer look at the phonon spectra of conventional superconductors like niobium and lead using inelastic neutron scattering, the answers would be quite discouraging. First of all, there exists the famous BCS theory of Bardeen et al. [2], which explains nearly all aspects of conventional superconductivity. Second, the worldwide interest is oriented towards high temperature superconductivity in cuprates, heavy Fermions and the pnictides. Thus the first experiments of this thesis, which addressed the phonon linewidths of superconducting niobium and lead, were only intended as a short test-bed of the resolution properties of the new high-resolution neutron spectrometer at the research reactor FRM-2.

The phonon linewidth is proportional to the electron–phonon coupling parameter λ, which is an essential parameter describing the formation of Cooper pairs in phonon mediated superconductors. The superconducting energy gap, whose magnitude, symmetry, and temperature dependence are intimately related to Cooper pairing, can also be directly determined in phonon linewidth measurements. The opening of the gap results in a redistribution of electronic states and excitations in the immediate vicinity of the Fermi surface. Electron–phonon scattering is suppressed for phonon energies below the gap $2\Delta(T)$ due to the stability of the Cooper pairs below T_c. Discontinuities in the phonon linewidths are thus expected when the phonon energy exceeds $2\Delta(T)$. Consequently, the gap and its momentum dependent anisotropy (and also the pairing symmetry) can be accurately resolved from a map of these discontinuities in different crystallographic directions. While phonon energies are highly sensitive to the superconducting energy gap, their momenta can serve as a similarly comprehensive probe of the geometry of the Fermi surface, which also leaves an imprint on the phonon linewidth. Phonons which connect nearly parallel segments of the Fermi surface exhibit an enhanced electron–phonon scattering probability, and thus linewidth extrema (termed Kohn anomalies) are expected. A full image of the underlying Fermi surface is contained in a map of these Kohn anomalies.

The phonon-linewidth spectra of these long-known superconductors presented in this thesis showed new and unexpected features that were not visible in previous low resolution experiments. Anomalies were observed at phonon energies corresponding to the magnitude of the superconducting gap, 2Δ, in the electron spectrum. These features were not only visible, as expected, below the superconducting transition temperature T_c, but persist to much higher temperatures. A detailed analysis showed that these features originate from previously unknown Kohn anomalies. As these anomalies were observed both in niobium and lead, which have different crystal structures (BCC and FCC), Fermi surfaces, and energy gaps, it is likely that this link between 2Δ and the Kohn anomalies is more a general phenomenon. Thus the question arose whether the Kohn anomalies impose a limit on the magnitude of the 2Δ gap. The major part of this thesis was accordingly concentrated on the relation between Kohn anomalies and 2Δ. First, similar measurements were carried out along different high symmetry directions. Yet in all cases, the 2Δ gap was found to coincide with the lowest-energy Kohn anomaly indicating that this phenomenon cannot be attributed to an accident. Since the energies of the Kohn anomalies vary (within $\approx \pm 10\%$) for different crystallographic directions, this "locking" of 2Δ to the Kohn anomaly provides a simple explanation to the long quest for the origin of the gap anisotropy, which was already inferred from tunneling experiments in the 1960s and intensively discussed in the following decades.

To shed more light on this lock-in mechanism, other metallic superconductors were explored. A key candidate was the Pb–Bi alloy. Bi adds electrons and increases the radius of the Fermi surface. It was known from previous tunneling experiments that both the gap magnitude 2Δ and T_c increase with Bi concentration. Inelastic neutron spectroscopy on different Pb–Bi alloys revealed a shift of the Kohn anomalies to higher energies with increasing Bi content in lockstep with 2Δ, supporting the lock-in hypothesis. Preliminary data on elemental tantalum and the transition metal dichalcogenide $NbSe_2$ are also consistent with this hypothesis. An explanation of these data requires an extension of the accepted theoretical framework for conventional superconductivity.

Phonon linewidth experiments were also performed on the high temperature superconductor $La_{2-x}Sr_xCuO_4$, whose understanding is still in its infancy. Here the two dimensional Fermi surface should lead to strong features in the phonon spectrum induced by the electron–phonon interaction. The experiments were difficult due to the large number of phonon branches, which often introduce additional background and obscure the results. Nevertheless, the electron–phonon linewidths of transverse acoustic phonon branches in underdoped, optimally doped, and overdoped samples could be extracted from the data.

This thesis is divided in the following way. In Chaps. 2, 3, and 4, an experimental and theoretical overview of the electron–phonon interaction, and its importance in conventional and unconventional superconductivity is given. Chapter 5 presents the new generation inelastic neutron scattering spectroscopy, which combines the TAS and the NRSE techniques to attain an energy resolution of almost two orders of magnitude better than conventional neutron spectroscopy. Chapter 6 deals with the details of the NRSE-TAS resolution function and data analysis. The experimental results are presented and discussed in Chap. 7. Final conclusions are drawn in Chap. 8.

References

1. W. Meissner et al., Naturwissenschaften **21**, 787 (1933)
2. J. Bardeen et al., Phys. Rev. **108**, 1175 (1957)
3. L.N. Cooper, Phys. Rev. **104**, 1189 (1956)
4. J.G. Bednorz, K.A. Muller, Z. Phys. B **64**, 189 (1986)
5. H. Ding et al., Phys. Rev. B **54**, R9678 (1996)
6. M. Weger et al., Z Phys. B **101**, 573 (1996)
7. X.J. Zhou et al., Phys. Rev. Lett. **95**, 227002 (2005)
8. A. Lanzara et al., Nature **412**, 510 (2001)
9. D Reznik et al., Nature **440**, 1170 (2006)
10. J Lee et al., Nature **442**, 546 (2006)
11. T.P. Devereaux et al., Phys. Rev. Lett. **93**, 117004 (2004)
12. G. Sangiovanni et al., Phys. Rev. Lett. **97**, 046404 (2006)
13. Y. Kamihara et al., J. Am. Chem. Soc. **130**, 3296 (2008)
14. H. Ding et al., Europhys. Lett. **83**, 47001 (2008)
15. S. Tsuda et al., Phys. C **412**, 36 (2004)
16. T. Yildirim et al., Phys. Rev. Lett. **87**, 037001 (2001)
17. J.P. Carbotte, *Anisotropy Effects in Superconductivity*, ed. by H. W. Weber (Plenum, New York, 1977)
18. A.J. Bennett, Phys. Rev. **140**, A1902 (1965)
19. P.G. Tomlinson, Phys. Rev. B **13**, 4738 (1976)
20. N. Bock, D. Coffey, Phys. Rev. B **75**, 174513 (2007)
21. A. Floris, Phys. Rev. B **75**, 054508 (2007)
22. A.W. Overhauser, L.L. Daemen, Phys. Rev. Lett. **61**, 1885 (1988)
23. J.D. Short, J.P. Wolfe, Phys. Rev. Lett. **85**, 5198 (2000)

24. P.B. Allen, Phys. Rev. B **6**, 2577 (1972)
25. G. Shirane et al., *Neutron Scattering with a Triple-Axis Spectrometer*. (Cambridge University press, Cambridge, 1996)
26. F. Mezei, Z. Phys. **255**, 146 (1972)
27. R. Gähler et al., J. Phys. **49**, 1195 (1988)
28. T. Keller et al., Appl. Phys. A **74**, 127 (2002)

Chapter 2
Phonons and Their Interactions

2.1 The Phonon–Phonon Interaction

Within the harmonic approximation, phonons are non-interacting and have an infinite lifetime. Including higher terms (anharmonic terms) in the expansion of the potential leads to an interaction between phonons. As a result, a phonon from a given state defined by the wave vector $\mathbf{q}$ and the branch j of the dispersion spectrum $\omega_j(\mathbf{q})$ will decay into other phonons after a finite time. Phonon–phonon interactions involve different number of phonons in the interaction process.

In perturbation theory, the crystal potential is expanded as a power of displacement and the Hamiltonian may be written as:

$$H = H_o + \lambda H_3 + \lambda^2 H_4 + \lambda^3 H_5 + \cdots \tag{2.1}$$

where H_o is the harmonic Hamiltonian and H_3, H_4, H_5,... are the perturbation terms involving three, four, five ... interacting phonons. The simplest case is the three-phonon interaction where a phonon decays to form two other phonons and vice-versa. The possible interactions are shown in Fig. 2.1. The Hamiltonian for three-phonon processes reads[1]:

$$\begin{aligned} H_3 = & \frac{1}{3!} \sum_{'\mathbf{q}''} \sum_{jj'j''} \frac{\hbar^{3/2}}{2^{3/2} N^{1/2}} \frac{\phi(\mathbf{q}j, 'j', ''j'')}{\sqrt{\omega_j \omega_{\mathbf{q}'j'} \omega_{''j''}}} \delta_{\mathbf{q}+'+\mathbf{q}'',\mathbf{G}} \\ & \times (a^{\dagger}_{-j} + a_{\mathbf{q}j})(a^{\dagger}_{-'j'} + a_{'j'})(a^{\dagger}_{-''j''} + a_{''j''}) \end{aligned} \tag{2.2}$$

where $\phi(\mathbf{q}j, \mathbf{q}'j', \mathbf{q}''j'')$ is the 3×3 matrix element, $\mathbf{q}$, $\mathbf{q}'$, $\mathbf{q}''$ are the wave vectors of the three phonons involved in the process and $\mathbf{G}$ is the reciprocal lattice vector. The δ-function guarantees momentum conservation in these processes. The last

[1] For a derivation of the Hamiltonian refer to [1, 2].

P. Aynajian, *Electron–Phonon Interaction in Conventional and Unconventional Superconductors,* Springer Theses, DOI: 10.1007/978-3-642-14968-9_2,

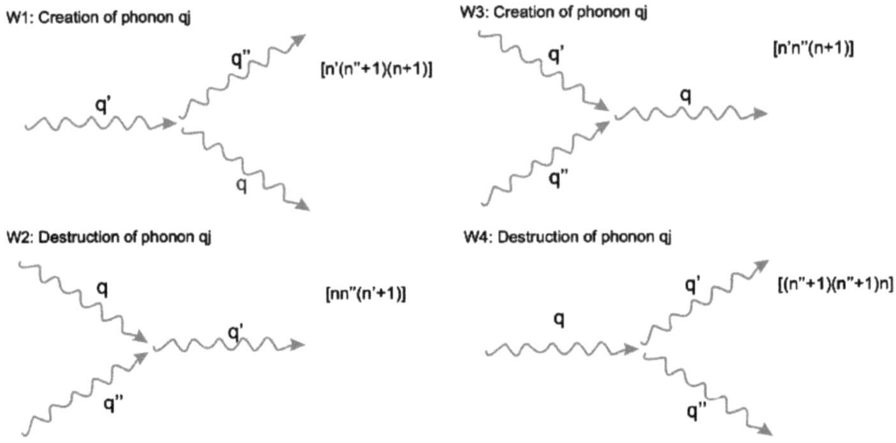

Fig. 2.1 Three-phonon interaction processes that may destroy or create the phonon $\mathbf{q}j$. Creation of a phonon in the process contributes to the transition probability by a factor $[(\hbar/2m\omega)(n+1)]^{1/2}$ while a phonon destruction contributes by $[(\hbar/2m\omega)(n)]^{1/2}$

three terms of Eq. 2.2 represent the different processes of creation and annihilation of the three phonons.

The probability per unit time, that a phonon in an initial state $|i\rangle$ with index $\mathbf{q}j$ will decay, as a result of one of the processes mentioned, into some other state $|f\rangle$ is given by the Fermi "Golden rule":

$$W(i \rightarrow f) = \frac{2\pi}{\hbar}|\langle f|H_3|i\rangle|^2 D_f(E) \tag{2.3}$$

where $D_f(E)$ is the density of the final states and H_3 is the three-phonon Hamiltonian of Eq. 2.2. The lifetime of this phonon $\mathbf{q}j$ is then defined as the reciprocal value of the decay rate. This rate is the sum of the rates of the processes where the phonon $\mathbf{q}j$ is created, minus the rate of the processes where the phonon $\mathbf{q}j$ is destroyed. Altogether it follows that the lifetime $\tau(\mathbf{q}j)$ is given by:

$$\begin{aligned} \frac{1}{\tau(\mathbf{q}j)} = \frac{\pi}{16N} \sum_{\mathbf{q}'j',\mathbf{q}''j''} & \frac{|\phi(\mathbf{q}j,\mathbf{q}'j',\mathbf{q}''j'')|^2}{\omega_{\mathbf{q}j}\omega_{\mathbf{q}'j'}\omega_{\mathbf{q}''j''}} \delta_{\mathbf{q}+\mathbf{q}'+\mathbf{q}',\mathbf{G}} \\ & \times \{[n'(n''+1)(n+1) - nn''(n'+1)] \times \delta(\omega_{\mathbf{q}} - \omega_{\mathbf{q}'} + \omega_{\mathbf{q}''}) \\ & + [n'n''(n+1) - n(n''+1)(n'+1)] \times \delta(\omega_{\mathbf{q}} - \omega_{\mathbf{q}'} - \omega_{\mathbf{q}''})\} \end{aligned} \tag{2.4}$$

Each partial probability contains the square of the transition matrix element times a delta function which guarantees energy and momentum conservation. The squares of the matrix elements for the different processes differ only in the occupation numbers of the phonons n, n', and n''. Higher order phonon processes, involving four or more phonons, are more numerous and arise from higher orders of the perturbation theory as seen from Eq. 2.1. Nevertheless the strength for these

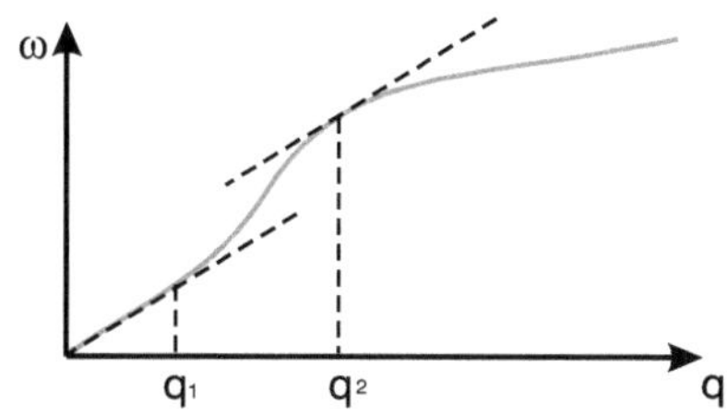

Fig. 2.2 Anomalous phonon dispersion, as is the case for liquid helium. Within the range $q_1 < q < q_2$, the phonon phase velocity, v_p exceeds the sound velocity v_s and spontaneous decay is allowed. $v_s = \omega/q$ as $q \rightarrow 0$; $v_p = \mathrm{d}\omega/\mathrm{d}q$

interactions decreases as the number of phonons in the interaction process increases.

As the temperature tends to zero, due to the absence of thermal phonon population (n' and $n'' = 0$), the decay probabilities for up-conversion processes vanish. The intrinsic phonon lifetime (due to *p–p* interaction) is thus dominated by spontaneous down-conversion decay towards lower energy phonons (Fig. 2.1d). In the case of a normal phonon dispersion, it has been discussed that four-phonon interaction is the lowest order decay process [3]. Three-phonon decay processes are kinematically not allowed. For anomalous phonon dispersion, however, where the phonon phase velocity exceeds the sound velocity, collinear (i.e. wave vectors of the involved phonons are all along the same direction) intra-branch three-phonon decay processes for transverse phonons become kinematically allowed. This leads to a finite phonon lifetime even at zero temperature. The phonon spectrum of superfluid helium exhibits such an anomaly, and spontaneous phonon decay at very low temperatures is experimentally observed for phonons with $v_g > v_s$ [4, 5] (v_g and v_s are the group velocity and the sound velocity, respectively). Figure 2.2 shows such an anomalous phonon dispersion. Within the range $q_1 < q < q_2$, the phonon phase velocity, v_p, exceeds the sound velocity. Only phonons within this range can decay to lower-energy phonons and therefore exhibit a finite lifetime at zero temperature.

2.2 The Electron–Phonon Interaction

In the band model description, electrons in a solid are quasi-particles which occupy one-electron states. They are described by Bloch functions $|\mathbf{k}, \sigma\rangle$, where $\mathbf{k}$ is the wave vector of the electron and σ is the spin.

In a perfect crystal, an electron propagates without scattering, however, the perfect periodicity is destroyed by the lattice vibrations of the atoms. These vibrations cause the electrons to have a certain probability of being scattered.

The electron–phonon interaction process induces the annihilation or creation of a phonon ($\mathbf{q}$, j) and a simultaneous excitation or de-excitation of the electron from

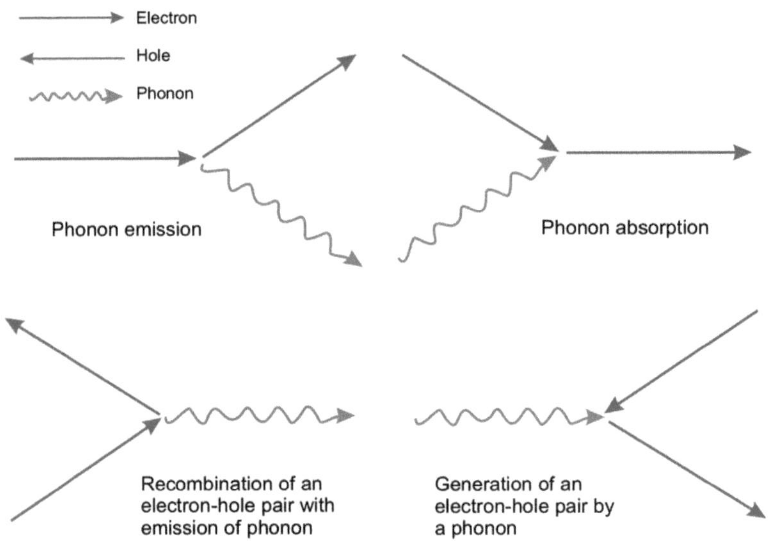

Fig. 2.3 The different possible electron–phonon interaction processes

state $|\mathbf{k}, \sigma\rangle$ to $|\mathbf{k} \pm \mathbf{q}, \sigma\rangle$. These two processes are illustrated in the top row of Fig. 2.3. Another two possible processes are illustrated in the second row of Fig. 2.3: recombination of an electron–hole pair with the creation of a phonon, and the creation of an electron–hole pair by the annihilation of a phonon. These four basic processes can be described quantum mechanically by a first-order perturbation calculation. The Hamiltonian of the electron–phonon interaction is[2]:

$$H_{e-\mathrm{p}} = \sum_{\mathbf{k},\mathbf{q}j} g(\mathbf{k}_1, \mathbf{k}_2; \mathbf{q}j) c^{\dagger}_{\mathbf{k}_1 j} c_{\mathbf{k}_2 j} (a^{\dagger}_{-\mathbf{q}j} + a_{\mathbf{q}j}) \tag{2.5}$$

where $c^{\dagger}_{\mathbf{k}_1 j}$ and $c_{\mathbf{k}_2 j} c_{\mathbf{k}2j}$ are the creation and annihilation operators for the quasi-particles with wave vectors $\mathbf{k}_1 = \mathbf{k} + \mathbf{q}$ and $\mathbf{k}_2 = \mathbf{k}$, respectively; $a^{\dagger}_{\mathbf{q}j}$ and $a_{\mathbf{q}j}$ are the creation and the annihilation operators of the phonon of energy $\omega_{\mathbf{q}j}$ and wave vector $\mathbf{q}$. The latter two operators in the Hamiltonian mean that two interactions are possible, one in which a phonon with wave vector $\mathbf{q}$ in branch j is created, and a second in which a phonon $\mathbf{q}, j$ is annihilated. Both processes are accompanied by an electron transition from an initial state $\mathbf{k}_1$ into a final state $\mathbf{k}_2$. The matrix element $g(\mathbf{k}_1, \mathbf{k}_2; \mathbf{q}j)$ describes the electron–phonon coupling and is defined as:

$$g(\mathbf{k}_1, \mathbf{k}_2; \mathbf{q}j) = -i\varepsilon(\mathbf{q}, j) \cdot \mathbf{q} V(\mathbf{q}) \tag{2.6}$$

[2] For more details and derivation of the Hamiltonian we point to the corresponding chapters in the books by Ziman [1], Grimvall [6], and Reissland [2].

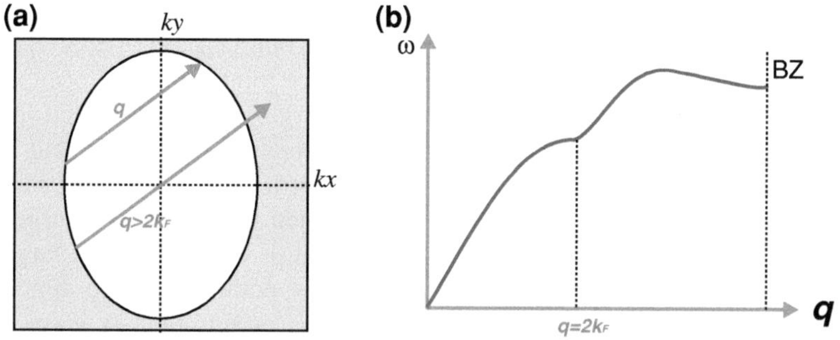

Fig. 2.4 **a** Schematic 2D Fermi surface. The *arrows* correspond to phonon wave vectors. **b** Phonon dispersion along **q**. The *kink* corresponds to the Kohn anomaly at $q = 2k_F$

where $V(\mathbf{q})$ is the matrix element of the electron–phonon pseudopotential and $\varepsilon(\mathbf{q}, j)$ is the phonon polarisation. Due to the factor $\varepsilon(\mathbf{q}, j) \cdot \mathbf{q}$, transverse phonons in the first Brillouin zone will not interact with the electrons. However in higher Brillouin zones, due to *Umklapp* processes the factor $\varepsilon(\mathbf{q}, j) \cdot \mathbf{q} \neq 0$ in general.

2.2.1 Kohn Anomalies in the Phonon Dispersion of Metals

The coupling function (Eq. 2.6) refers to the scattering of a quasi-particle from a point $\mathbf{k}_1$ to a point $\mathbf{k}_2$ in momentum space with $\mathbf{q} = \mathbf{k}_1 - \mathbf{k}_2$. Energy and momentum conservation require that both $\mathbf{k}_1$ and $\mathbf{k}_2$ lie on the Fermi surface.[3] This immediately introduces a restriction on the phonon wave vector **q**: phonon wave vectors connecting nested parts of the Fermi surface will strongly interact with the electrons leading to a large phonon damping, whereas those which do not span the Fermi surface will not interact with the electrons. It has been pointed out by Kohn [7] that the interaction of phonons with the conduction electrons in a metal should cause anomalies in the phonon spectra. The phonon dispersion should exhibit kinks at wave vectors $\mathbf{q} + \mathbf{G} = 2\mathbf{k}_F$, where $\mathbf{k}_F$ is the Fermi wave vector and **G** is the reciprocal lattice vector. Figure 2.4a illustrates a schematic two-dimensional (2D) Fermi surface. Phonons with wave vector $\mathbf{q} + \mathbf{G} < 2\mathbf{k}_F$ (represented by the short arrow in Fig. 2.4a) can always excite quasi-particle quasi-hole pair since they span the Fermi surface. Their self-energy is therefore renormalized with respect to the bare phonon energy. For $\mathbf{q} + \mathbf{G} > 2\mathbf{k}_F$ this condition is not fulfilled and those phonons do not interact with the electrons. At $\mathbf{q} + \mathbf{G} = 2\mathbf{k}_F$ a discontinuity results in the momentum dependence of the electron–phonon interaction. This is reflected

[3] This is due to the fact that the typical phonon energy is in the meV range which is three orders of magnitude smaller that the typical electron energy.

in the phonon dispersion illustrated schematically in Fig. 2.4b. The anomaly at $\mathbf{q} + \mathbf{G} = 2\mathbf{k}_F$ is the Kohn anomaly after Walter Kohn [7].

The strength of the Kohn anomaly depends on the joint density of occupied and unoccupied electronic states. If their quantity is large, conduction electrons may, become unstable with respect to a spatially inhomogeneous perturbation. To first order, the response of the electrons to such a perturbation is measured by the generalized susceptibility $\chi_{\mathbf{q}}$. Instability sets in when this quantity diverges. This happens in a nesting situation, that is, in a situation in which there are large areas of the Fermi surface which are parallel or nearly parallel. In two-dimensional systems, such instabilities usually lead to a charge- or spin-density wave ground state.

Highly accurate neutron measurements of $\omega(\mathbf{q})$ are required to reveal these anomalies. Such measurements were performed on various metals and indicate a structure of singularities that is consistent with the Fermi surface geometry. Figure 2.5 shows the phonon dispersion in elemental Pb along the $(\xi\xi0)L$ branch [8] and the $(\xi\xi0)T$ branches (measured on TRISP). Kohn anomalies can be seen in the dispersion. The locations of those anomalies correspond to diameter of the Fermi surface [10]. Similar Kohn anomalies have also been observed in the phonon dispersion of transition-metal dichalcogenides 2H-TaSe$_2$ and 2H-NbSe$_2$ [9]. Pronounced softening were detected by inelastic neutron scattering at $q = 0.2$ in both systems due to the quasi-two-dimensional nature of their Fermi surface which leads to a charge density wave ground state.

Figure 2.6 shows the phonon dispersion of the $(\xi\xi\xi)L$ branch in Pb$_{1-x}$Tl$_x$ alloys [11]. Pure Pb exhibits a strong Kohn anomaly at $q \approx 0.42$ rlu. By doping Pb with

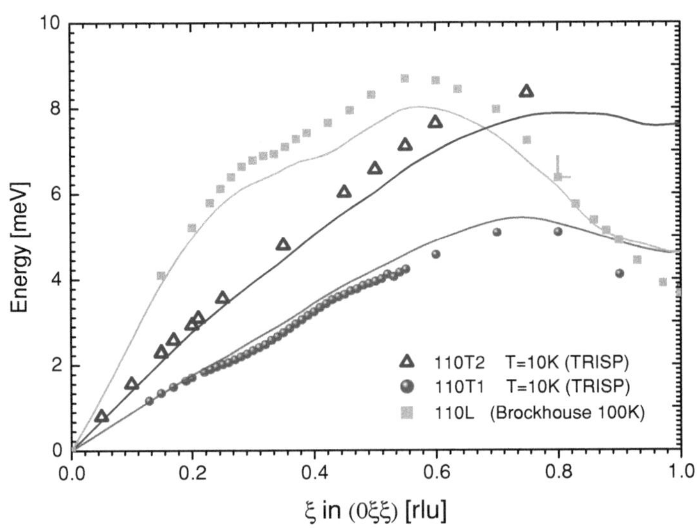

Fig. 2.5 Phonon dispersion of Pb along the ($\xi\xi$ 0) branch. Kohn anomalies are observed which are consistent with the Fermi surface geometry. The transverse branches (T$_1$ and T$_2$) were measured on TRIPS whereas the data for longitudinal branch is taken from Brockhouse [8]

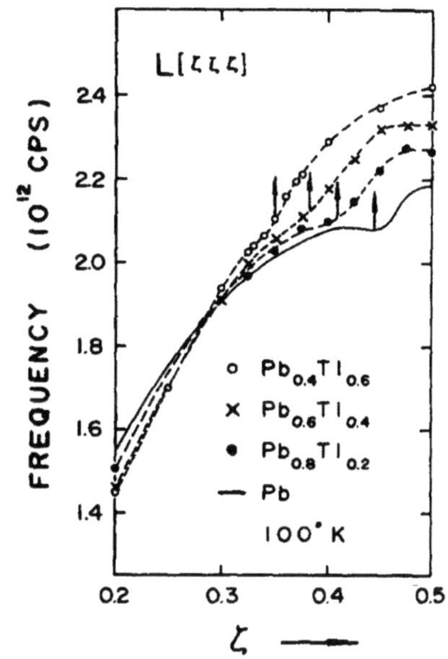

Fig. 2.6 Dispersion curves for $Pb_{1-x}Tl_x$ alloys along the $(\xi\xi\xi)L$ branch. The lines are guide to the eye. The *arrows* indicate the location of the Kohn anomalies which, via doping Pb with Tl, progressively move to lower momenta due to change of the Fermi surface. Figure is taken from Ng et al. [11]

Tl (hole doping since Tl has one fewer electron), the Fermi surface shrinks. This is reflected in the location of the Kohn anomaly, which gradually moves to smaller wave vectors.

In the above figures, the phonon dispersion was plotted along a specific direction. In two or three dimensions, the locus of Kohn anomalies form a surface, the Kohn surface whose shape and strength reflect the geometry of the Fermi surface. For example, from Fig. 2.4a, the Kohn anomaly along k_x lies at a smaller wave vector than that along the k_y direction. The strengths of the anomalies also differ due to the different curvatures along those directions.

References

1. J.M. Ziman, *Electrons and Phonons.* (Oxford, London, 1962)
2. J.A. Reissland, *The Physics of Phonons.* (Wiley, Toronto, 1972)
3. A. Berke, J. Phys. C Solid State Phys. **21**, 2205 (1988)
4. H.J. Maris et al., Phys. Rev. Lett. **25**, 220 (1970)
5. L.P. Pitayevski et al., Phys. Rev. B **14**, 263 (1976)
6. G. Grimvall, *The Electrons–Phonons Interaction in Metals*, vol. XVI. (North Holland, 1980)
7. W. Kohn, Phys. Rev. Lett. **2**, 393 (1959)
8. B.N. Brockhouse, *Inelastic Scattering of Neutrons in Solids and Liquids.* (IAEA, Vienna, 1961)
9. D.E. Moncton et al., Phys. Rev. B **16**, 801 (1977)
10. R. Stedman et al., Phys. Rev. **163**, 567 (1967)
11. S.C. Ng et al., Solid State Comm. **5**, 79 (1967)

Chapter 3
Conventional Superconductivity

3.1 Overview of the Microscopic Theory of Superconductivity

3.1.1 The BCS Formalism

The microscopic theory of superconductivity in metals was developed by Bardeen et al. (BCS) [1] almost 50 years after the first observation of superconductivity in Hg in 1911. BCS outlined that the presence of an attractive interaction between electrons leads to an instability of the normal electronic state and the formation of a coherent many-body ground state. This attractive potential binds two electrons of opposite spin together forming the so called Cooper pairs with zero total spin, which condense to a single statet and form a bosonic condensate.

Until around 1950, it was a mystery as to what could be the mechanism for superconductivity. The standard description of electrons in a metal was Bloch's picture. Obviously, one had to find some interaction between the electrons which lowers the total energy of the system when the electrons form a superconducting state.

The mechanism behind the weak attractive force binding the Cooper pairs was first suggested by Fröhlich [2]. He proposed that the electron–phonon interaction leads to superconductivity. This hypothesis of a pairing mechanism driven by the electron–phonon interaction was confirmed by experiments that showed that the critical temperature, T_c, varied with isotopic mass.

When the Hamiltonian, $H_{e\text{–}p}$, is written in the form described in the previous chapter Eq. 2.5, it is not transparent that it gives rise to an attraction between two electrons. There, a phonon is created or annihilated in a scattering process with an electron. However, after a phonon is, e.g. created, it can be annihilated by another scattering process with an electron. Thus, a phonon is exchanged by two electrons giving rise to an effective interaction (see Fig. 3.1). Such a process can occur in second and in higher orders in the electron–phonon interaction. Without stating the

P. Aynajian, *Electron–Phonon Interaction in Conventional and Unconventional Superconductors,* Springer Theses, DOI: 10.1007/978-3-642-14968-9_3,

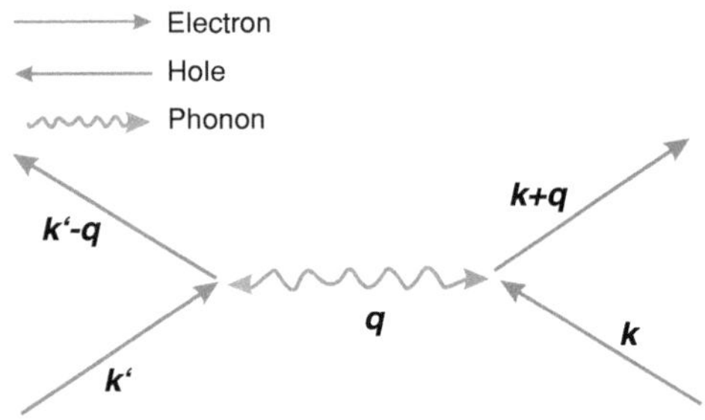

Fig. 3.1 Diagrammatic representation of the effective electron–electron interaction through the exchange of a phonon q

mathematical details here, Fröhlich showed indeed that the Hamiltonian which contains the effective interaction can be written as [3]:

$$H_{e-e} = \sum_{q,k,k'} V_{kq} c^{\dagger}_{k'} c_{k'+q} c^{\dagger}_{k-q} c^{\dagger}_{k} \tag{3.1}$$

with

$$V_{kq} = \frac{|\, g(\mathbf{k}_1, \mathbf{k}_2; \mathbf{q}j) \,|^2 \, \hbar\omega_q}{(E_{k_1} - E_{k_2})^2 - \hbar^2 \omega_q^2} \tag{3.2}$$

It is immediately seen that for a small region around the Fermi energy with $|\,(E_{k_1} - E_{k_2})\,| < \hbar\omega_q$, phonons mediate an attractive interaction among electrons. One could then roughly argue that a new bound state consisting of two fermions could appear, such that the bound pair would have bosonic character, and hence, similarly to ^4He, a super fluid phase could appear. Since in this case the pair would be charged, superconductivity would take place.

That such a bound state can occur even in the presence of many electrons, is the subject of the theory developed by BCS [1, 4]. The BCS theory allows the calculation of several features such as the transition temperature to superconductivity and the opening of a gap in the one-particle excitation spectrum. In their theory, the following approximations were made:

a. The pairing interaction is weak.
b. The density of states is not varying too fast near the Fermi surface.
c. The pairing interaction is constant and independent of momentum ($V_{qk} = V_o$) within the cutoffs $\pm\hbar\omega_D$ near the Fermi surface, where ω_D is the Debye frequency, and zero otherwise.

These approximations lead to the BCS equations for the transition temperature T_c and the superconducting gap Δ[1]:

[1] For details, the reader is advised to refer to [1].

$$T_c = 1.14\hbar\omega_D \exp\left(-\frac{1}{N(E_F)V_o}\right) \tag{3.3}$$

and

$$\Delta = 2\hbar\omega_D \exp\left(-\frac{1}{N(E_F)V_o}\right) \tag{3.4}$$

where $N(E_F)$ is the electronic density of states at the Fermi surface. Since the pairing interaction is constant, the gap has an isotropic s-wave symmetry.

3.1.2 The Eliashberg Formalism

The equilibrium superconducting properties of any material can be derived from the knowledge of the spectral function or the Eliashberg function $\alpha^2F(\omega)$ where α is the average electron–phonon interaction and $F(\omega)$ is the phonon density of states.

$$\alpha^2F(\omega) = \frac{1}{N(E_F)}\sum_{qj}\delta(\omega-\omega_{qj})\sum_{k} \mid g(\mathbf{k}_1,\mathbf{k}_2;\mathbf{q}j) \mid^2 \times \delta(E_{k_1}-E_F)\delta(E_{k_2}-E_F) \tag{3.5}$$

This function measures the contribution of phonons with frequency ω to scattering processes of electrons at the Fermi level and is related to the dimensionless electron–phonon coupling parameter λ,

$$\lambda = 2\int_0^{\omega_o} \frac{\alpha^2F(\omega)}{\omega}d\omega = \sum_{qj}\lambda_{qj} \tag{3.6}$$

which corresponds roughly to the product $N(E_F)V_o$ of the original BCS formula Eq. 3.12.

Within the Elliasberg theory, the superconducting gap is given by:

$$\Delta = 2\hbar\omega_c \exp\left(\frac{\lambda-\mu*}{\lambda+1}\right) \tag{3.7}$$

and

$$T_c = 1.14\hbar\omega_c \exp\left(\frac{\lambda-\mu*}{\lambda+1}\right) \tag{3.8}$$

where $\hbar\omega_c$ is a cutoff frequency related to the phonon density of states and μ^* is the Coulomb pseudopotential which is the reduced Coulomb repulsion experienced by a Cooper pair. This equation has essentially summarized all the detailed information contained in the e–p spectral function $\alpha^2F(\omega)$ into two parameters, λ and ω_c. However depending on the material, μ^* can offset the effect of λ. An important conclusion from Eqs. 3.7 and 3.8 is that they depend exponentially on material-specific parameters. Even though this framework provides a remarkably successful description of the physical properties of conventional low-temperature superconductors, a reliable calculation of the transition temperature and the energy gap from first principles still presents a formidable challenge to theory.

3.2 Experimental Approach to the Electron–Phonon Interaction and Superconductivity

Measurements of electron–phonon interaction in elemental superconductors have been mostly limited to tunnelling experiments [5–8]. From these experiments, $\alpha^2F(\omega)$ can be deduced in favorable cases, and if $F(\omega)$ is known from other measurements[2] then α^2 can be obtained. Unfortunately, in addition to this being an indirect method to measure the electron–phonon interaction, the results are often dependent on the junction preparation.

In 1972 Allen [11] showed that the coupling parameter λ_{qj} is related to the electron–phonon linewidth Γ_q^{e-p}, and is given by

$$\Gamma_{qj}^{e-p} = \pi N(E_F)\hbar\omega_{qj}^2\lambda_{qj} \tag{3.9}$$

where λ_{qj} is the contribution to λ from the particular phonon qj. Thus λ_{qj} can be determined experimentally if we can measure the electron–phonon contribution to the phonon linewidth. Equation 3.9 offers an ideal tool for the experimental study of the electron–phonon interaction.

As has been discussed in the previous chapter, the phonon linewidth contains a contribution from the anharmonic phonon–phonon interaction, $\Gamma = \Gamma_{e-p} + \Gamma_{p-p}$. As Γ_{p-p} is strongly temperature dependent and Γ_{e-p} is roughly temperature independent,[3] it is possible to extract Γ_{e-p} by measuring $\Gamma(T)$. In addition to intrinsic effects however, an artificial linewidth broadening in inelastic neutron spectroscopy rises from instrumental resolution.

Measurements of the phonon linewidth in superconductors offer a direct way to extract the electron–phonon part Γ_{e-p}. In the superconducting phase it is useful to distinguish between two mechanisms for phonon damping. One is due to phonons scattering from thermally excited quasiparticles and vanishes as $T \longrightarrow 0$. The other process involves the direct excitation of the quasiparticles across the superconducting energy gap by phonons. Since the minimum electronic excitation energy in a superconductor is $2\Delta(T)$, all phonons with energy $\hbar\omega_{\mathbf{q}j} \leq 2\Delta(T)$ are

[2] e.g. from neutron scattering measurements.

[3] except in the vicinity of the superconducting transition temperature.

energetically incapable of exciting an electron–hole pair and thus have a vanishing $\Gamma_{e\text{-}p}$ when $T \ll T_c$. The linewidth Γ in this case is limited by the instrumental resolution and possible *p–p* contributions. By subtracting the linewidths measured at $T \ll T_c$ from those just above T_c, one obtains $\Gamma_{e\text{-}p}$. The change in the phonon–phonon linewidth within this temperature range is negligible.

There are measurements of this type by Axe and Shirane on Nb_3Sn [10, 12] and by Shapiro et al. on Nb [9] by inelastic neutron scattering on a triple axis spectrometer. Abrupt changes in the linewidths as a function of temperature near T_c were observed along the $[\xi\xi0]$ T_2 direction in Nb while for Nb_3Sn changes were observed along the $[\xi\xi0]$ T_1 direction. The linewidth $2\Gamma_{e\text{-}p}$ was about 0.07 meV for Nb and 0.8 meV for Nb_3Sn. The huge difference in $2\Gamma_{e\text{-}p}$ between these two materials can be understood from Eq. 3.9. $2\Gamma_{e\text{-}p}$ depends on the coupling parameter, λ, and on the electronic density of states $N(E_F)$. These two quantities are much larger in Nb_3Sn compared to Nb.

In addition to the direct measurement of the momentum dependent *e–p* interaction, phonon linewidth measurements allow the determination of the magnitude and the temperature dependence of the superconducting gap. In principle, measurements of $\Gamma_{e\text{-}p}$ along different momentum directions can even probe the anisotropy of the superconducting energy gap. In practice however, it is difficult to obtain the requisite energy resolution of a few μeV, comparable to the normal state *e–p* linewidth $\Gamma_{e\text{-}p}$. Such investigations have therefore been limited to a few selected materials (such as Nb and Nb_3Sn). Attempts to resolve the *e–p* linewidths in Pb, among other materials, by conventional neutron spectroscopy have been unsuccessful [13, 14].

In Chap. 5 a neutron spin echo method will be introduced that has sufficient energy resolution to study in detail the *e–p* interaction and detect fine anomalies in the phonon spectrum of superconductors.

3.3 Anisotropy of the Superconducting Energy Gap

In the original BCS formulation, a constant effective *e–e* interaction and spherical Fermi surface were assumed. The energy gap equation, when solved under these assumptions, yielded a momentum independent isotropic solution. These assumptions are, of course, rather schematic, and experimentally, gap anisotropy has been observed in most superconductors. These measurements indicate a range of energy gaps that vary with angle and may be different on different sheets of the Fermi surface [15–19].

The anisotropy of the gap is directly related to the anisotropy of V_o which includes both the anisotropy of the Fermi surface, as well as the anisotropy of the phonon spectrum. The anisotropy within the BCS model can be simply seen if V_o is written in an anisotropic form as:

$$V_{kk'} = V_o[1 + a(k)][1 + a(k')] \tag{3.10}$$

where k and k' specify points on a spherical Fermi surface and $a(k)$ is the anisotropy term which is a function of momentum direction only. Then clearly the gap will be momentum dependent exhibiting the same anisotropy as $V_{kk'}$:

$$\Delta_k = \Delta_o(1 + a_k) \tag{3.11}$$

Similarly one can solve the T_c equation to obtain:

$$T_c = 1.14\hbar\omega_D \exp\left[-\frac{1}{N(E_F)V_o(1 + a_k^2)}\right] \tag{3.12}$$

This last equation demonstrates that T_c is increased by anisotropy. However scattering due to impurities washes out the gap anisotropy and reduces T_c. It is important to note that in such a model, no attempt is made to relate the effective potential $V_{kk'}$ to the fundamental parameters of the metal.

To calculate the anisotropic energy gap for a specific material at various points on the Fermi surface, one needs the knowledge of the momentum dependence of the spectral function $\alpha_k^2 F_k(\omega)$, where k refers to a particular electronic state on the Fermi surface. The Fermi surface average of $\alpha_k^2 F_k(\omega)$ gives the isotropic function (Eq. 3.5) which refers to all electrons participating in superconductivity.

Different anisotropic sources lead to an anisotropy of the function $\alpha_k^2 F_k(\omega)$.

- *The Fermi surface*: intersection of a spherical Fermi surface with the Bragg planes results in significant distortions. These deviations from sphericity provide a major source of gap anisotropy.
- *The electronic wave function and e–p interaction*: the electronic wave function can differ from a plane wave specially in regions where the Fermi surface is not spherical. Since this wave function enters as the initial state in the electron–phonon matrix element $g(\mathbf{k}_1, \mathbf{k}_2; \mathbf{q}j)$ (see Eq. 3.5), anisotropy will arise in $\alpha_k^2 F_k(\omega)$ from this source.
- *The phonon spectrum*: the phonon energies in general depend on the direction of $\mathbf{k}$ in the first Brillouin zone and therefore will also be a source of anisotropy in $\alpha_k^2 F_k(\omega)$.
- *Umklapp processes*: for a definite electronic transition from $\mathbf{k}_1$ to $\mathbf{k}_2$, the momentum transfer $\mathbf{q} = \mathbf{k}_1 - \mathbf{k}_2$ serves to label the phonon mode involved. For each initial state $\mathbf{k}_1$, different phonon modes are involved. Not only the set of phonons are different but also the reciprocal lattice vectors involved in the Umklapp processes are quite distinct. These difference can lead to large anisotropies.

How big can the anisotropy in conventional superconductors be?

Figure 3.2 represents the calculated results of the gap anisotropy in pure Al (after [20, 21]). Panel a shows the spectral function $\alpha_k^2 F_k(\omega)$ along the three high symmetry directions, (100), (110), and (111). The curves were calculated assuming a free electron model. The considerable amount of anisotropy is mainly due only to Umklapp anisotropy and the anisotropy of the phonon spectrum itself. Panel b shows the calculated $\lambda_k \equiv \lambda(\theta, \phi)$ along three constant ϕ arcs and panel c

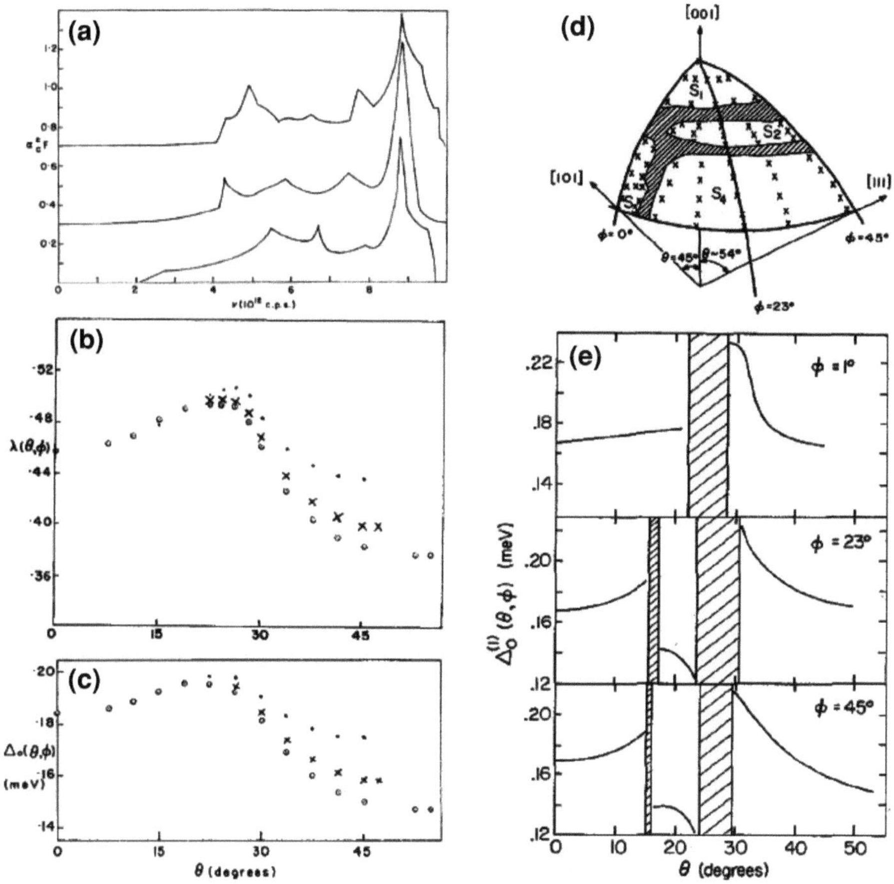

Fig. 3.2 **a** Calculated $\alpha_k^2 F_k(\omega)$ versus phonon frequency for the three high symmetry directions in Al. The *three curves* are displaced vertically from each other to facilitate comparison. The *lower curve* is for the (100) direction, the *middle*, (110) direction, and the *upper*, (111) direction. **b** The directional electron–phonon mass-enhancement parameter, $\lambda(\theta, \phi)$, for Al calculated in the free electron model for *phi* = 1° (*dotted*), 23° (*multiplication symbol*), and 45° (*open circle*), as a function of θ. **c** The energy gap $\Delta(\theta, \phi)$ in the free electron model (same notations as in **b**). **d** A schematic diagram showing the relative positions in the 1/48th irreducible zone on the real Fermi surface of the 62 points at which calculations have been performed. **e** The directional energy gap edge $\Delta(\theta, \phi)$ including anisotropy due to phonon spectrum, the *e–p* Umklapp processes, and electronic band structure.The *shaded regions* indicate places where there is no Fermi surface. The figure is taken from [20] and [21]

shows the results for the Fermi surface variation of the gap in Al (Note that band structure corrections have been left out. The Fermi surface is assumed to be spherical and the electronic wave functions to be plane waves). A comparison with the results for $\lambda(\theta, \phi)$ shows that the gap varies in much the same way. From panel c, the authors calculated the mean square anisotropy parameter $a^2 = \langle a_k^2 \rangle$ (see Eq. 3.12),

$$a^2 = \frac{\langle \Delta_k^2 \rangle - \langle \Delta_k \rangle^2}{\langle \Delta_k \rangle^2} \tag{3.13}$$

and found $a^2 = 0.0084$ which is small but significant, concluding that phonon spectrum anisotropy and Umklapp anisotropy provide a significant source of variation in $\Delta(\theta, \phi)$.

In the above results, a spherical Fermi surface was assumed. The authors also studied the modifications in $\Delta(\theta, \phi)$ by introducing band structure effects. Figure 3.2d shows a projection of the Al Fermi surface onto a sphere. The shaded area indicates places where no Fermi surface exist. These are the segments where the Bragg planes intersect the Fermi surface. In these regions, the Fermi surface can be quite distorted and the multiple plane wave nature of the electronic wave functions becomes important. Figure 3.2e shows the gap anisotropy along different directions including all the different sources of anisotropy in the calculation. The mean square anisotropy parameter calculated from panel e is $a^2 = 0.0187$, which is more than twice the result found for the free electron model. It is thus evident that band structure effects make a significant contribution to the gap anisotropy in Al. Another interesting feature arising from gap anisotropy is the enhancement of T_c (see Eq. 3.12). The authors calculated the ratio $T_c^p/T_c^d = 1.091$, where T_c^p and T_c^d are the superconducting transition temperatures in the pure and dirty limit, respectively. The theory predicts 9% increase in T_c compared to a 5% increase found in experiments [20].

A similar sort of calculation for the gap anisotropy in the strong coupling superconductor Pb has also been performed [22]. The value of the mean square anisotropy for Pb is found to be $a^2 = 0.009$, which is much smaller than that for Al. Recent first-principles density functional calculations performed in Pb yield a superconducting gap which exhibits largely different values on the two sheets of the Fermi surface [23]. In these calculations, the value of the gap is underestimated ($1.5\text{ meV} < 2\Delta < 2.2\text{ meV}$), but the anisotropy value (30%) is larger than in previous theories.

How good is the agreement between the available theories and experimental results? From the experimental point of view, there are two different methods to measure the anisotropy of the energy gap: those from which only the existence of anisotropy can be deduced, and those which can identify selective processes that allow a direct determination of $\Delta(\mathbf{k})$ for various **k**-directions in the crystal. In the first case, the anisotropy is determined from macroscopic properties such as the specific heat, change in T_c due to impurities, and the critical magnetic field. These properties are determined by an angular average of $\Delta(\mathbf{k})$ over all states **k** on the Fermi surface. They are directly proportional to the parameter $\langle a^2 \rangle$. On the other hand, experiments such as superconductive tunneling, ultrasonic attenuation, and microwave absorption can directly access the momentum variation of the gap. A brief review of the gap anisotropy measurements from the various mentioned techniques is given in Table 3.1. The results suggest the existence of gap anisotropy in many materials. There is, however, a large disagreement between values obtained by different techniques [16]. Comparison of experimental and theoretical results is thus severely limited by experimental difficulties [24, 25]

Table 3.1 Summary of the gap anisotropy parameter $\langle a^2 \rangle$ for various materials measured by various techniques

Technique	Al	In	Pb	Sn
Electromagnetic radiation	0.02	0.01	0.001/0.0002	0.02/0.01
Ultrasonic attenuation	0.011/0.02	0.0003/0.01	0.0009	0.04/0.03
Specific heat	0.01	0.01	0.33??	0.0035
Critical magnetic field	0.013	0.027	0	0.022
Thermal conductivity	0	0.0008	0	0.01
Tunneling	0.009/0.001	–	0.007/0.031	0.026

The table is taken from [16]

which may arise from the difference in bulk and surface sensitive techniques, uncontrolled momentum–space averaging in tunneling spectroscopy [18, 19, 24–26], and non equilibrium effects in phonon imaging experiments [27].

It is therefore still largely unclear to what extent the calculations and the experimental results are reliable.

References

1. J. Bardeen et al., Phys. Rev. **108**, 1175 (1957)
2. H. Fröhlich, Phys. Rev. **79**, 845 (1950)
3. H. Fröohlich, Proc. R. Soc. A **215**, 291 (1952)
4. J Bardeen, Physics **7**, 54 (1972)
5. WL McMillan et al., Phys. Rev. Lett. **14**, 108 (1964)
6. JM Rowell, Phys. Rev. Lett. **10**, 334 (1963)
7. GI Rochlin, Phys. Rev. **153**, 513 (1966)
8. I Giaever, Phys. Rev. Lett. **5**, 464 (1960)
9. S. Shapiro et al., Phys. Rev. B **12**, 4899 (1975)
10. J.D. Axe et al., Phys. Rev. B **8**, 1965 (1973)
11. P.B. Allen. Phys. Rev. B **6**, 2577 (1972)
12. J.D. Axe et al., Phys. Rev. Lett. **30**, 214 (1972)
13. A. Furrer et al., Phys. Statue Solidi **42**, 821 (1970)
14. R. Youngbloud et al., Solid State Commun. **27**, 1433 (1978)
15. S. Tsuda et al., Phys. C **412**, 36 (2004)
16. J. Bostock et al. (Plenum, New York, 1977)
17. T. Yokoya et al., Science **294**, 2518 (2001)
18. P.L. Richards, Phys. Rev. Lett. **7**, 412 (1961)
19. K. Yoshihiro et al., J. Phys. Soc. Jpn **28**, 262 (1970)
20. C.R. Leavens et al., Ann. Phys. **70**, 338 (1972)
21. H.K. Leung et al., J. Low Temp. Phys. **24**, 25 (1976)
22. P.G. Tomlinson et al., Phys. Rev. B **13**, 4738 (1976)
23. A. Floris et al., Phys. Rev. B **75**, 054508 (2007)
24. B.L. Blackford et al., Phys. Rev. **186**, 397 (1969)
25. G.I. Rochlin, Phys. Rev. **153**, 513 (1967)
26. B. Farnworth et al., Phys. Rev. B **10**, 2799 (1974)
27. T.L. Head et al., Phys. Rev. B **78**, 054516 (2008)

Chapter 4
Unconventional Superconductivity

4.1 Phenomenology of Cuprates

Superconductivity in the cuprates has been a problem standing for more than two decades and there is still no agreement on the pairing mechanism. For some recent reviews see [1–3]. Superconductivity, however, is not the only interesting property of cuprates. The normal state ($T > T_c$) has also anomalous properties. Still today it is a challenge to explain the normal state properties and in particular the underdoped region of the phase diagram, which remains a controversial topic.

The basic building blocks of superconducting cuprates are the CuO_2 planes separated by layers with rare-earth elements. From a chemical point of view, $La_{2-x}Sr_xCuO_4$ (LSCO) is among the most simple cuprates. Holes are doped into the CuO_2 plane by substituting La with Sr. The hole doping content is therefore directly determined by the concentration of Strontium. The lattice structure of LSCO is shown in Fig. 4.1a. It exhibits a tetragonal or orthorhombic structure depending on doping and temperature. The second order transition from low-temperature orthorhombic (LTO) to high-temperature tetragonal (HTT) is usually used to estimate the doping level of the samples. The lattice constants, at room temperature, are $a = b = 3.8\,\text{Å}$ and $c = 13.2\,\text{Å}$ (tetragonal notation).

A schematic diagram of the Fermi surface of LSCO as extracted from angle resolved photoemission spectroscopy (ARPES) [4] is shown in Fig. 4.1b for different hole concentrations. As the hole concentration decreases, the Fermi surface near (π, 0) smoothly moves through (π, 0) so that the topological center of the Fermi surface is turned over from (0,0) to (π, π) for 20% Sr concentration [4]. On the other hand, the position of the Fermi surface near ($\pi/2$, $\pi/2$) is less sensitively dependent on the hole concentration.

High-T_c superconductors (HTSC) exhibit a rich phase diagram. Figure 4.2 illustrates schematically the different phases as a function of the hole doping. This section is devoted to a brief examination of the generic phase diagram of HTSC.

P. Aynajian, *Electron–Phonon Interaction in Conventional and Unconventional Superconductors*, Springer Theses, DOI: 10.1007/978-3-642-14968-9_4,

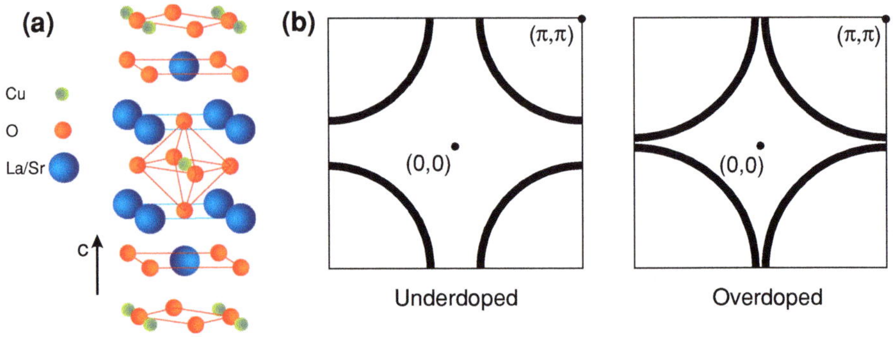

Fig. 4.1 **a** Crystal structure of $La_{2-x}Sr_xCuO_4$. The CuO_2 planes are seen in the bottom and top of the unit cell. Sr substitution for La, changes the carrier concentration on the CuO_2 planes, drastically altering the electronic properties of this compound. **b** Schematic diagram of the Fermi surface of $La_{2-x}Sr_xCuO_4$ in the underdoped and the overdoped phases

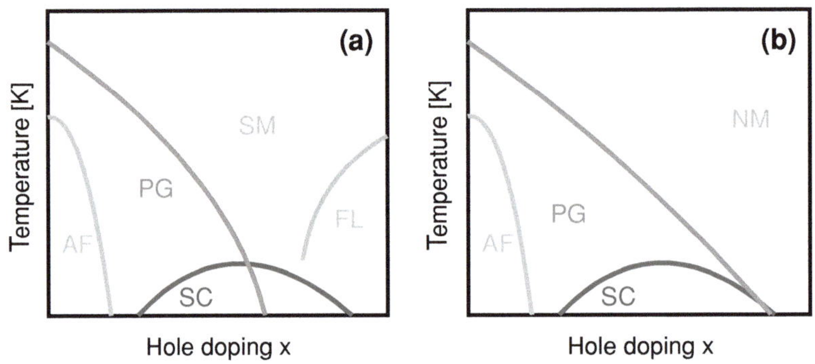

Fig. 4.2 **a** and **b** are two different ways to draw the phase diagram. The overdoped region is a normal metal (Fermi liquid) and the undoped compound is an antiferromagnetically ordered Mott insulator. Between these extrema superconductivity exists. The controversy about the pseudogap phase leads people to draw the phase diagram in two different ways as shown in the panel (**a**) and (**b**)

4.1.1 Insulating Regime

Band calculations predict a half-filled metallic band in the undoped phase of the cuprate superconductors. The parent compounds are however Mott insulators [5] due to the strong Coulomb interaction U between the d-shell Cu electrons. The energy cost t for an electron to jump to a neighboring site is much smaller than the Coulomb interaction ($t \ll U$) and therefore the electrons prefer to localize on the Cu sites. In this insulating regime the Cu spins interact via a superexchange interaction that causes them to order antiferromagnetically below the Neel temperature $T_N \approx 300\,\text{K}$.

4.1.2 Metallic Regime

In the other extreme ($t \gg U$) the electrons delocalize and form a metallic regime. Electrons in metals have been successfully described by Landau's Fermi liquid theory. A signature of a Fermi liquid is a quadratic temperature dependence of the in-plane resistivity $\rho_{ab}(T) = \rho_o + AT^2$ where A is a constant and the T^2 term stems from electron–electron interaction. At low-temperatures ($T < 50$ K) such a T-dependence has been observed in the overdoped and non-superconducting region ($x > 0.3$) of LSCO [6]. Often empirical observations also suggest that conventional Fermi liquid theory applies to the overdoped regime of HTSC.

4.1.3 Strange Metal Phase

It is still debated to what extent Fermi liquid theory applies to the normal state of HTSC in general. The normal state of BCS superconductors is a Fermi liquid and the BCS theory identifies superconductivity as an instability of the Fermi sea. The normal state of optimally doped HTSCs is, however, not a standard Fermi liquid since it exhibits anomalous metallic behavior. For example the in-plane resistivity scales linearly with temperature $\rho_{ab}(T) = \rho_o + AT$ for $T_c < T < 1,000$ K [7]. This T-dependence demonstrates unambiguously that the normal state of optimal doped HTSCs is not a conventional Fermi liquid. The normal state around optimal doping is therefore often denoted as a strange metal. The microscopic mechanism for this strange metal is still controversial.

4.1.4 Pseudogap

Another long standing problem is the normal state of underdoped cuprates. This region of the phase diagram is often denoted as the pseudogap (PG) phase. ARPES has directly confirmed the existence of a pseudogap in the normal state ($T > T_c$) electronic spectra of the underdoped cuprates [8]. Near the (π, 0) point a strong suppression of the density of states has been observed. The pseudogap phase has been studied directly and indirectly by a vast number of experimental techniques [9]. This has led to a number of different definitions of pseudogap transition temperature T_{PG} and the pseudogap magnitude Δ_{PG}. Although most experimental techniques agree that T_{PG} and Δ_{PG} increase with decreasing doping in the underdoped regime [8], the pseudogap phase remains an unsettled issue.

Broadly speaking there exist two different interpretations of the pseudogap phase. The first idea is that the pseudogap is a precursor to superconductivity. In this picture, preformed pairs form locally but without long range phase coherence in the normal state below the crossover temperature T_{PG}. This view has been supported by the observation of vortex-motions through the Nernst effect in the normal state of underdoped cuprates [10, 11]. As a precursor for superconductivity the pseudogap phase should exist above the superconducting dome. The phase diagram is therefore often drawn as shown in Fig. 4.2a. The other class of models argues that the pseudogap phase is related to a generic thermodynamic transition into a phase with a hidden order parameter. Within this scenario, the order vanishes inside the superconducting dome as shown in Fig. 4.2b. The topology of this phase diagram is similar to that of certain heavy fermion compounds where the superconducting dome appears around an antiferromagnetic quantum critical point [12]. In heavy fermion materials superconductivity and non-Fermi-liquid-behavior often appear in conjunction with a quantum critical point [13]. This might suggest that there may be some aspects of quantum critical physics that are shared between the heavy fermion and cuprate materials [14].

4.2 Proposed Theories of High T_c Superconductors

There still exists no consensus about the microscopic mechanism of high-temperature superconductivity. A microscopic theory should not only explain superconductivity, but describe the phase diagram in general. It should provide an explanation for the strange metal phase and the pseudogap. At zero hole doping the model should correctly describe the Mott insulating phase and in the strongly overdoped regime a Fermi liquid like phase should appear. Many ideas have been proposed during the last couple of decades. Here, a few of the most influential theories are outlined. For a recent discussion of the various scenarios, see [15].

4.2.1 Phonon-Mediated HTSC

Some researchers emphasize the role of lattice-vibrations [1, 16], suggesting that HTSC is mediated by phonons as in conventional BCS superconductors. For some doping levels, an isotope effect on T_c has been observed. However the isotope effect vanishes at optimum doping. In undoped cuprates, an isotope effect may arise from competition between superconductivity and phases with localized electrons (e.g. stripe phases). It is difficult to reproduce the experimentally observed d-wave pairing symmetry in such models.

4.2.2 RVB Approach

Very early on, Anderson suggested the concept of a resonant-valence bond (RVB) state in which magnetic singlet pairs exist in a spin liquid like state [17]. RVB correlations can lead to various other ground states such as antiferromagnetism and superconductivity. The RVB concept provides an explanation for d-wave superconductivity and pseudogap state, where singlet pairs exist but phase coherent superconductivity is absent. The RVB-approach is also able to describe the strange metal phase [18] without the use of a quantum critical point. This model provides an explanation to many of the salient features in the phase diagram of HTSCs. For a more detailed review see [1, 19].

4.2.3 Spin Fluctuations

While the RVB approach treats the problem with a large Coulomb repulsion U, the spin fluctuation models start with a weak interaction U [20]. These models also provide a natural explanation for d-wave symmetry of the gap function. A naive way to understand these models is to assume that the spin fluctuations play the same role as the phonons do for the conventional superconductors. These models are also able to explain the strange metal phase since an antiferromagnetic quantum critical point will lead to a marginal Fermi liquid like spectrum [21].

4.2.4 Inhomogeneity-Induced Pairing

Another class of theories emphasizes the tendency of these materials to phase separate into hole rich and hole poor regions. The fluctuations of these mesoscopic structure are then responsible for phase coherent superconductivity. In the overdoped region the system is more homogeneous, which weakens the pairing. On the other hand, in the underdoped region the system is too granular and phase coherence can therefore not be obtained. In these models that involve stripes or phase separation, the pseudogap is a phase where local pairing is possible. For a review see [22].

4.3 Experimental Overview of the Normal and Superconducting Gaps in Cuprates

The symmetry of the superconducting order parameter is a forward step in understanding the mechanism behind high temperature superconductivity. Conventional BCS superconductors exhibit an s-wave order parameter, reflecting the spherically symmetric nature of the pair wave function. For cuprates, theoretical analysis of the CuO_2 plane with consideration of the strong on-site Coulomb interaction leads to other symmetries of the order parameter [23–26]. In particular, pairing theories based on the Hubbard model lead to a d-wave order parameter or a mixed symmetry order parameter with a strong d-wave component [23–26].

The magnitude and symmetry of the gap function, $\Delta(\mathbf{k})$, can be determined experimentally. ARPES has played an important role in understanding the cuprates. It provides the ability to measure the magnitude of the superconducting gap as a function of momentum. High resolution ARPES experiments are consistent with the d-wave nature of the superconducting gap $2\Delta(\mathbf{k}) = \Delta[\cos(k_x a) - \cos(k_y a)]$ with a zero (node) gap along the diagonal at 45° to the CuO_2 bond direction and a maximum gap (antinode) along the $(\pi, 0)$ direction [27].

The gap magnitude ($\Delta_{max} \approx 30\,\text{meV}$) in the HTSCs is much larger than that predicted by BCS. Although different techniques yielded slightly different gap values, all are in the range of $2\Delta/k_B T_c = 6$ to 12, much larger than the value 4.3 given by BCS. This is not the main difference between BCS and HTSC. In conventional superconductors, the appearance of the superconducting energy gap in the electronic spectrum indicates pairing of electrons into Cooper pairs, a phase coherence, and a simultaneous transition into a macroscopic superconducting state. In HTSC the situation is more complicated. A pseudogap (whose relation to the superconducting gap remains a mystery) develops well above T_c [9, 28]. Figure 4.3

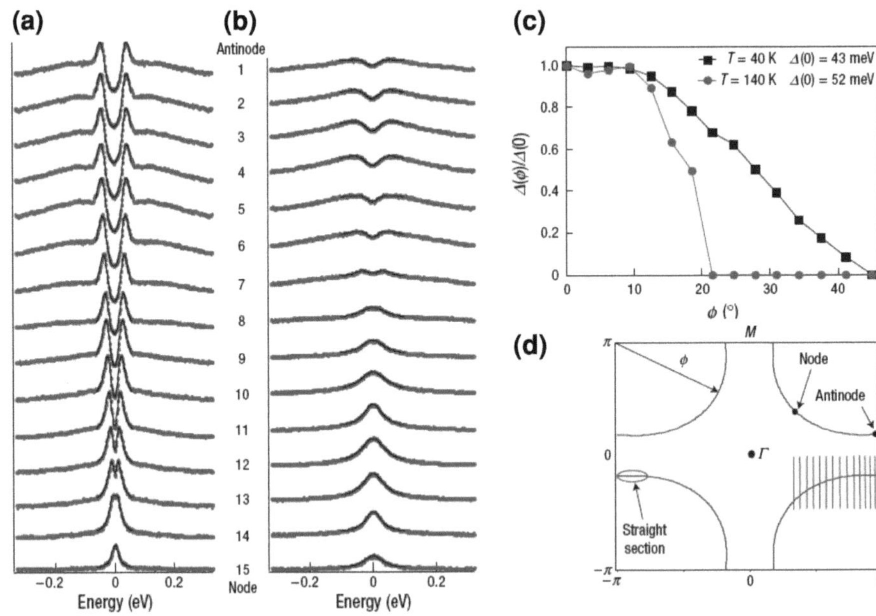

Fig. 4.3 Angle resolved photoemission spectroscopy on $Bi_2Sr_2CaCu_2O_8$ sample with the $T_c = 90\,\text{K}$ at $T = 40\,\text{K}$ in the superconducting state (a) and $T = 140\,\text{K}$ in the pseudogap state (**b**). The different energy distribution curves (EDCs) cover the whole Fermi surface from the node (*lowest curve*) up to the antinode (*uppermost curve*). **c** Variation of the gap around the Fermi surface extracted from (**a**) and (**b**). The uncertainty in the gap is ± 4meV for the pseudogap, and ± 2meV for the superconducting gap. **d** Location of the momentum cuts (*red lines*), Fermi surface (*blue curves*), and special points (node and antinode) in the zone. Figure from [29]

shows ARPES spectrum for the temperature and momentum evolution of the superconducting and the normal state gaps in $Bi_2Sr_2CaCu_2O_8$ with $T_c = 90K$ (taken from [29]). The spectra in panel (a) is in the superconducting state. One can clearly see the variation of the superconducting gap along the Fermi surface. It decreases monotonically from its maximum value at the antinode (top curve) eventually vanishing at the node (bottom curve). The spectra in panel (b), which is in the normal state above T_c, shows a normal state pseudogap persisting to temperatures far above T_c. A closer look at Fig. 4.3 shows that the momentum dependence of the gaps in the superconducting and the pseudogap states are dramatically different (panel c). This behavior has not been observed in previous ARPES measurements [30] because the nodal BCS-like gap region has been overlooked as a result of insufficient momentum-space sampling.

With the enhanced momentum resolution of ARPES measurements, distinct energy gaps has been reported by other authors as well [31]. Figure 4.4 shows high resolution ARPES spectra for $Bi_2Sr_2CaCu_2O_{8+\delta}$ in the underdoped and overdoped region of the phase diagram suggesting two distinct gaps. Close to the nodal

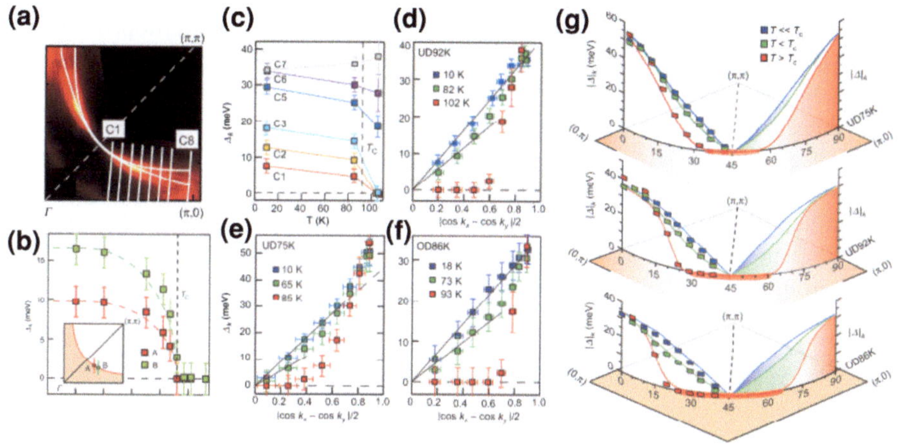

Fig. 4.4 **a** A partial Fermi-surface mapping measured at 102 K in a quadrant of the first Brillouin zone in slightly underdoped Bi2212 ($T_c = 92$ K). **b** Temperature dependence of the fitted gap size at Fermi-surface crossing points near the node, as indicated in the *insets*. The *dashed lines* show the temperature dependence of the superconducting gap based on weak-coupling BCS theory and serve as a guide to the eye. **c–f** Temperature and momentum dependence of the gap profile. **g** Schematic illustrations of the gap function evolution for three different doping levels: Underdoped sample with $T_c = 75$ K, Underdoped sample with $T_c = 92$ K, Overdoped sample with $T_c = 86$ K. At $10K$ above T_c there exists a gapless Fermi arc region near the node; a pseudogap has already fully developed near the antinodal region (*red curves*). With increasing doping, this gapless Fermi arc elongates (*thick red curve* on the Fermi surface), as the pseudogap effect weakens. At $T < T_c$ a d-wave like superconducting gap begins to open near the nodal region (*green curves*); however, the gap profile in the antinodal region deviates from the simple $d_{x^2-y^2}$ form. At a temperature well below T_c ($T \ll T_c$) the superconducting gap with the simple $d_{x^2-y^2}$ form eventually extends across entire Fermi surface (*blue curves*) in UD92K and OD86K but not in UD75K. The figure is obtained from Ref. [31]

direction, the superconducting gap opens at T_c and has a BCS-like temperature dependence (panel b). Close to the antinodal direction however, the gap persists above T_c (panel c). Figure 4.4g shows a schematic summary of the temperature dependent evolution of the gap function at three different doping levels. The normal state pseudogap strongly deviates from d-wave symmetry. The authors [31] suggest that the energy gap opening at T_c near the nodal region is associated with the order parameter of the superconducting state, whereas the pseudogap near the antinodal region represents an energy scale associated with a different mechanism that may or may not be related to superconductivity. The opening of the gap at T_c was difficult to understand in the context of previous ARPES [30, 32, 33] and scanning tunnelling microscope results dominated by the antinodal region.

Notably, the temperature dependence of the scanning tunnelling microscope (STM) spectrum has recently been revisited and the coexistence of two energy gaps in underdoped cuprates has also been suggested [34, 35]. In particular, a normalization procedure reveals that one of the gaps disappears at T_c [35].

Figure 4.5 shows representative spectra [34] measured at temperatures far above T_c that allow a comparison of the behavior of overdoped, optimal and underdoped $Bi_2Sr_2CaCu_2O_{8+\delta}$ samples. Once the pairing gaps collapse at high temperatures, the overdoped and optimally doped samples show remarkably similar electron–hole asymmetric spectra. In contrast, the underdoped sample (UD73) shows very different, V-shaped spectra with an ill-defined gap, which is insensitive to increasing temperature. These spectra are qualitatively different from those observed on optimal and overdoped samples above T_c: the gap closes in overdoped samples but 'fills in' for underdoped samples.

A key question is whether the gap above T_c is associated with pairing, and what determines the temperature at which incoherent pairs form. Spatially resolved STM measurements of gap formation on samples with different T_c values show that pairing gaps nucleate in nanoscale regions above T_c. These regions proliferate

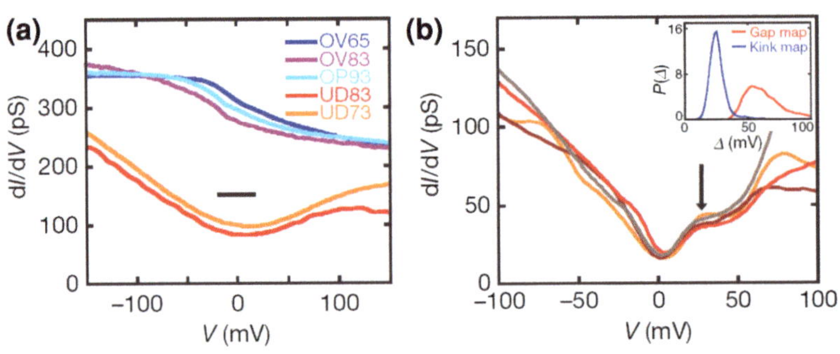

Fig. 4.5 Spatially resolved STM measurements of gap formation in $Bi_2Sr_2CaCu_2O_{8+\delta}$ samples with different T_c values (hole concentration of 0.12 to 0.22). **a** Representative spectra taken at specific locations well above T_c for all doping levels. **b** Spectra taken at 20 K at various locations on the underdoped (UD73) sample. Within the large gap a kink is observed at lower energy, indicated by the *arrow*. *Inset*, distributions of the large gap and the kink energy on the UD73 sample at 20 K. The figure is from [34]

as the temperature is lowered, resulting in a spatial distribution of gap sizes in the superconducting state. Despite the inhomogeneity, every pairing gap develops locally at a temperature T_p slightly above T_c, following the relation $2\Delta/k_B T_p = 7.9 \pm 0.5$. This observation is also supported by measurements of fluctuating superconductivity, which have shown that the onset temperatures for these fluctuations are well below the pseudogap temperature for underdoped samples [36]. At very low doping however, all the different spectra indicate the presence of a 'kink' indicating the importance of a lower energy scale, as shown in Fig. 4.5b. The probability of observing such spectra in optimal or overdoped samples is negligible. The inset of Fig. 4.5b shows a mapping of the spatial variation and distribution of each energy scale. The presence of this additional energy scale indicates that the large gaps seen in underdoped samples cannot be simply associated with pairing.

Evidence for two energy scales, one related to pairing and one related to strong pseudogap behavior in underdoped copper oxides, has also been observed in recent Raman measurements. Electronic Raman spectroscopy is capable of probing quasiparticles of the superconducting state in selected parts of momentum space, namely the antinodal (B1g symmetry, cross polarised rotated 45° with respect to the CuO bond) and nodal (B2g symmetry, cross polarised parallel to the CuO bond) regions, where the amplitude of the superconducting gap reaches its maximum and vanishes respectively. Figure 4.6 shows the characteristic ratios $\omega_{\rm AN}/T_c^{\rm max}$ (Antinodal) and $\omega_N/T_c^{\rm max}$ (Nodal) for several families of cuprates as a function of doping at a fixed temperature well below T_c ($T_c^{\rm max}$ is T_c at optimal doping). The figure is taken from [37]. These ratios have a universal dependence on doping. For underdoped compounds, two distinct scales are present which are clearly separated with the two ratios behaving in opposing ways as a function of doping. A unique energy scale (and doping dependence) is recovered in the optimally doped and overdoped regime.

In summary, all data clearly demonstrates the existence of two distinct energy scales in the superconducting state of underdoped cuprates, with opposite doping dependence. A pseudogap formation owing to preformed Cooper pairs is not consistent with the observed results from various techniques. The distinct temperature and doping dependence seem to suggest a competing nature between the nodal BCS-like superconducting gap and the antinodal pseudogap. It is not unrealistic to assume that the pseudogap is due to the formation of a competing state, such as a density-wave state, which explains why some parts of the Fermi surface do not develop superconducting coherence.

4.4 Importance of the Electron–Phonon Interaction in HTSC

After the discovery of HTSC, there was initially much interest in the electron-phonon (*e–p*) interaction. It was, however, soon concluded that the *e–p* interaction

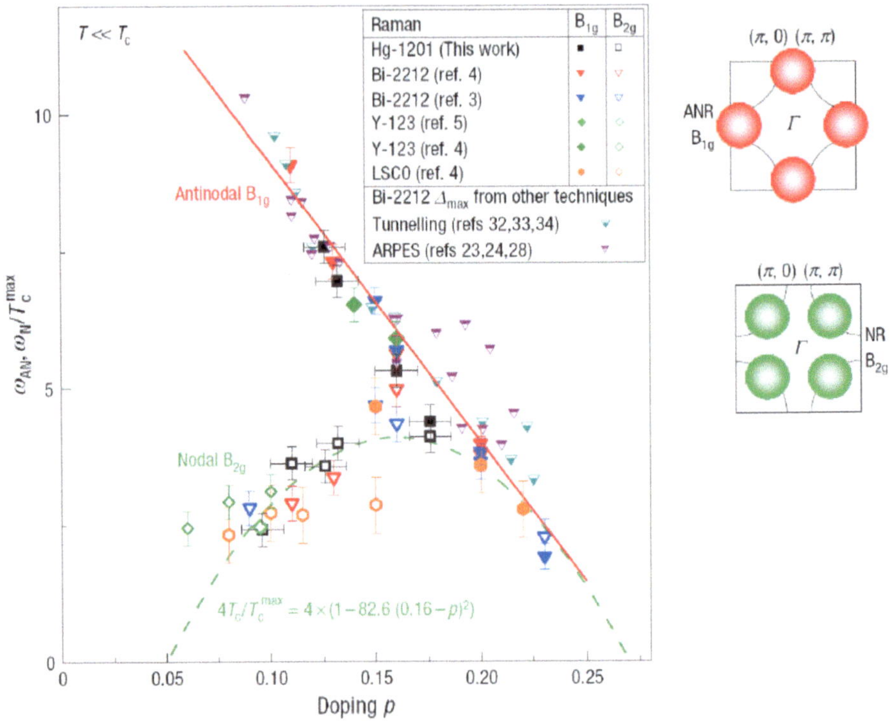

Fig. 4.6 Universal doping dependence of the ratios $\omega_{\mathrm{AN}}/T_c^{\max}$ and $\omega_N/T_c^{\max}$ of the antinodal (B1g) and nodal (B2g) superconducting peaks (obtained from [37]). The error bars on the B1g and B2g Raman peak locations have also been reported and show unambiguously that there are two energy scales in the underdoped side of hole-doped cuprates. The references in the figure are those in [37]

is too weak to explain superconductivity, in particular d-wave superconductivity. The interest was shifted to purely electronic models. Recently however, there has been substantial experimental evidence that *e–p* interaction plays an appreciable role for a number of properties.

There have been extensive studies of phonons in HTSC using neutron scattering. For reviews, see, e.g., Pintschovius [38]. Figure 4.7 shows the phonon dispersion for undoped (left-hand side) and doped (right-hand side) $La_{2-x}Sr_xCuO_4$ [39]. Shell model calculations give a rather accurate description of almost all phonon branches. One striking exception is the highest mode of Δ_1 symmetry for doped systems. This is the so-called half-breathing phonon, which is a bond-stretching vibration of the oxygen atoms in the CuO_2 plane. This phonon is rather well described by the shell model for the undoped system. However, doping leads to a strong softening half way along the $(\xi, 0, 0)$ direction. This softening is anomalous in the sense that it is not captured by the shell model (see Fig. 4.8 after [38]). The anomalous behavior is also illustrated by the large broadening up to 5 meV in the doped system. Both the softening and the width indicate that this

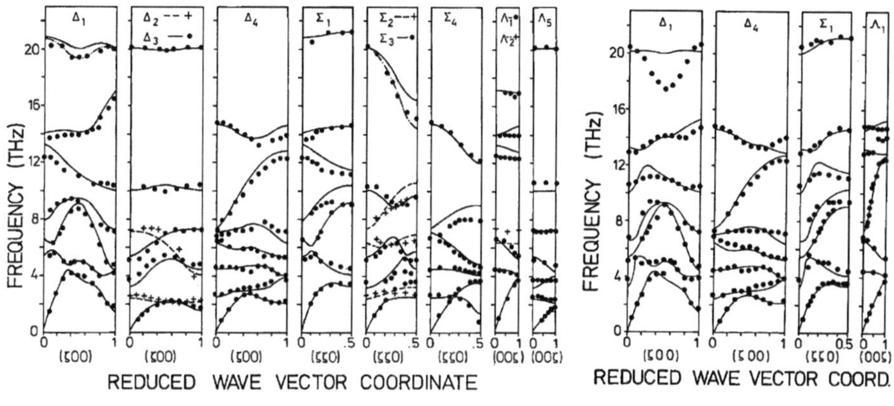

Fig. 4.7 Phonon dispersion curves of undoped La_2CuO_4 and underdoped $La_{1.9}Sr_{0.1}CuO_4$. The figure was taken from Chaplot et al. [38]

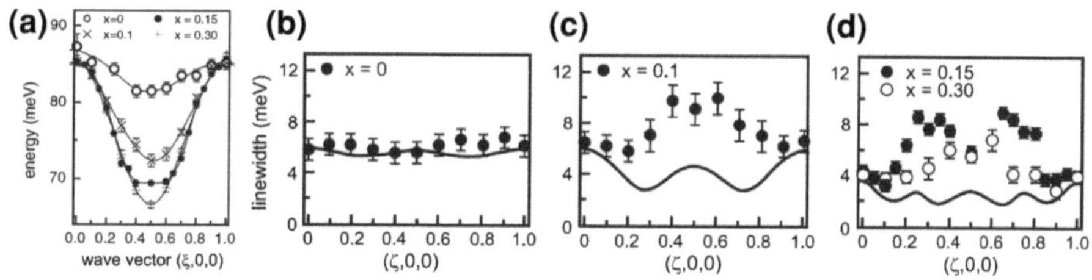

Fig. 4.8 **a** Dispersion of the longitudinal (Δ_1-symmetry) plane polarized Cu-O bond-stretching vibrations along the (ξ00)-direction in $La_{2\text{-}x}Sr_xCuO_4$ [2, 7]. Lines are a guide to the eye. Linewidth of the same phonon as a function of the reduced wavevector (ξ00) for doping $x = 0$ (**b**), $x = 0.1$ (**c**) and $x = 0.15$, ~ 0.30 (**d**). The *full line* shows the experimental resolution including focusing effects. The figure illustrates the large broadening for the doped system and the small intrinsic broadening for the undoped (x = 0) system. Figure after Pintschovius [38]

phonon couples strongly to doped holes. Similar anomalies are found in several cuprates and are particularly large for systems with static stripe order [40].

There has also been a substantial interest in a B1g phonon involving out-of-plane and out-of-phase c-axis vibrations of oxygen atoms in the CuO_2 plane. This phonon has an energy of about 42 meV for $YBa_2Cu_3O_{7\text{-}\delta}$ and has been studied using both Raman [41, 42] and neutron scattering [43–45]. This phonon shows an interesting change of frequency and linewidth as the compound is cooled below T_c (see Fig. 4.9) and an asymmetric Fano line shape is observed in Raman scattering.

Signatures of strong coupling of phonons to electrons have also been observed in ARPES experiments [28]. Most of the interest in the *e–p* interaction has been created by the observation of a kink in the experimental electron dispersion for several cuprates [46–50]. Some typical results are shown in Fig. 4.10 obtained from [46]. The authors emphasize that such a kink is found for three different

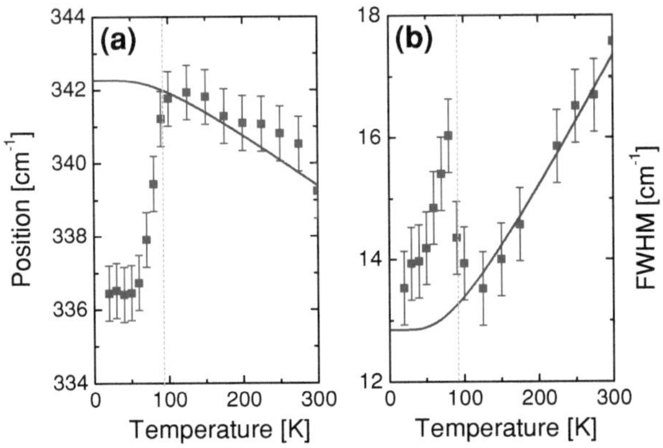

Fig. 4.9 Temperature dependence of frequency and the linewidth (FWHM) of the B1g-like mode of $YBCO_{6.95}$ ($T_c = 92K$) for the YY polarisation. The *blue line* corresponds to the temperature dependence of the pure phonon–phonon interaction (anharmonicity). Superconductivity induced changes are observed below T_c due to the opening of the superconducting gap. The figure is obtained from [41]

families of compounds (LSCO, Bi2201 and Bi2212), for different dopings and both below and above T_c. For noninteracting electrons, the ratio of the slopes below and above the kink is expected to be given by the dimensionless electron–phonon coupling $1 + \lambda'$. Figure 4.10f shows the change of slope $1 + \lambda'$ giving a value of $\lambda' \approx 1$.

While the early measurements showed only one kink at about 70 meV, later work found several structures at smaller binding energies [51, 52]. ARPES measurements by Zhou et al. in $La_{2\text{-}x}Sr_xCuO_4$ [51] shows structures in the real part of the self-energy at various energies (about 40–46 and 58–63 meV and possibly at 23–29 and 75–85 meV), suggesting that there is a coupling to bosons at these energies. The multiple features show marked difference from the magnetic excitation spectra measured in LSCO which is mostly featureless and doping dependent. In comparison, the features in [51] show more resemblance to the phonon density of states (DOS), measured from neutron scattering on LSCO, in the sense of the number of modes and their positions. This similarity between the extracted fine structure and the measured phonon features favors phonons as the nature of bosons involved in the coupling with electrons. In this case, in addition to the half-breathing mode at 70–80 meV that was previously considered strongly coupled to electrons (Fig. 4.8), those results suggest that several lower energy phonons are also actively involved.

In conclusion, evidence from different spectroscopic techniques has shown a substantial electron–phonon interaction in HTSC. It is therefore of particular interest to study linewidth broadening and energy shift of certain phonon modes as a function of momentum as well as doping by high resolution inelastic neutron scattering experiments.

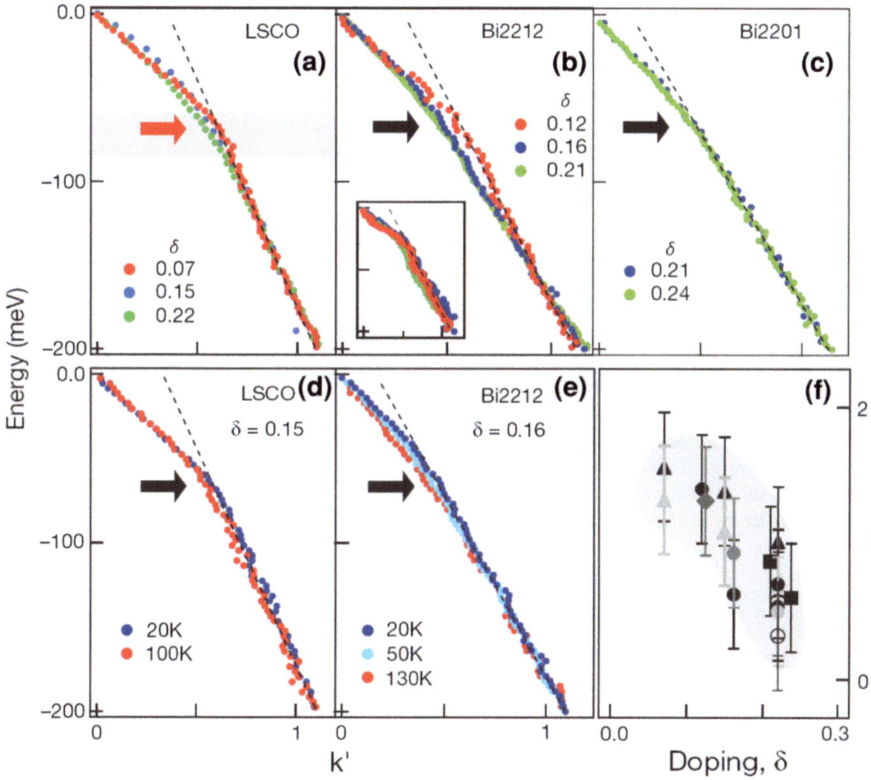

Fig. 4.10 Dispersion in the (0, 0)−(π, π) direction for LSCO, Bi2201 and Bi2212 for different dopings and temperatures. The results in (**a**) and (**b**) were obtained for T = 20 K and in (**c**) for T = 30 K. The kink position in (**a**) is compared to the half-breathing phonon energy at $q = (\pi, 0)$ (*thick red arrow*) and the phonon width and dispersion (*shaded area*). *Inset* in (**b**), dispersions off the (0, 0)−(π, π) direction, showing also a sharpening of the kink on moving away from the nodal direction. The *black arrows* indicate the position of the kink in the dispersions. **f** shows the change $1 + \lambda'$ of the slope at the kink. The figure is obtained from [46]

References

1. P.A. Lee et al., Rev. Mod. Phys. **78**, 17 (2006)
2. J. Orenstein et al., Science **288**, 468 (2000)
3. M.R. Norman, Rep. Prog. Phys., **66**, 1547 (2003)
4. A. Ino et al., Phys. Rev. B **65**, 094504 (2002)
5. M. Imada et al., Rev. Mod. Phys. **70**, 1039 (1998)
6. S. Nakamae et al., Phys. Rev. B **68**, 100502 (2003)
7. H. Takagi et al. Phys. Rev. Lett. **69**, 2975 (1992)
8. H. Ding et al., Nature **382**, 51 (1996)
9. T. Timusk et al., Rep. Prog. Phys. **62**, 61 (1999)
10. Y. Wang et al., Phys. Rev. B **64**, 224519 (2001)
11. Y. Wang et al., Phys. Rev. Lett. **88**, 257003 (2002)
12. N.D. Marthur et al., Nature **394**, 39 (1998)

13. H.V. Löhneysen et al., Rev. Mod. Phys. **79**, 1015 (2007)
14. P. Coleman et al., J. Phys. Condens. Matter **13**, R723 (2001)
15. Nat. Phys. **2**, 138 (2006)
16. N. Gedik et al., Science **316**, 425 (2007)
17. P.W. Anderson, Science **235**, 1196 (1987)
18. P.W. Anderson, Nature Phys. **2**, 626 (2006)
19. P.W. Anderson et al., J. Phys. Condens. Matter **16**, R755 (2004)
20. A.V. Chubukov et al., Phys. Rep. **288**, 355 (1997)
21. A. Abanov et al., Adv. Phys. **52**, 119 (2003)
22. S.A. Kivelson et al., Rev. Modern Phys. **75**, 1201 (2003)
23. D.J. Scalapino et al., Phys. Rev. B **34**, 8190 (1986)
24. N.E. Bickers et al., Phys. Rev. Lett. **62**, 961 (1989)
25. P. Monthoux et al., Phys. Rev. Lett. **67**, 3448 (1991)
26. N. Bulut et al., Phys. Rev. Lett. **68**, 706 (1992)
27. H. Ding et al., Phys. Rev. B **54**, R9678 (1996)
28. A. Damascelli et al., Rev. Mod. Phys. **75**, 473 (2003)
29. A. Kanigel, et al. Nature Phys. **2**, 447 (2006)
30. M.R. Norman et al., Nature **392**, 157 (1998)
31. W.S. Lee et al., Nature **450**, 81 (2007)
32. A.G. Loeser et al., Science **273**, 325 (1996)
33. A.G. Loeser et al., Phys. Rev. B **56**, 14185 (1997)
34. K. Gomes et al., Nature **447**, 569 (2007)
35. M.C. Boyer et al., Nature Phys. **3**, 802 (2007)
36. Z.A. Xu et al., Nature **406**, 486 (2000)
37. M. Le Tacon et al., Nature Phys. **2**, 537 (2006)
38. L. Pintschovius. Phys. Status Solidi B **242**, 30 (2005)
39. S.L. Chaplot et al., Phys. Rev. B **52**, 7230 (1995)
40. D. Reznik et al., Nature **440**, 1170 (2006)
41. M. Bakr et al., Phys. Rev. B (2009, submitted)
42. C. Thomsen et al., Phys. Rev. B **37**, 9860 (1988)
43. D. Reznik et al., Phys. Rev. B **78**, 094507 (2008)
44. D. Reznik et al., Phys. Rev. Lett. **75**, 2396 (1995)
45. H.F. Fong et al., Phys. Rev. Lett. **75**, 316 (1995)
46. A. Lanzara et al., Nature **412**, 510 (2001)
47. T. Valla et al., Science **285**, 2110 (1999)
48. A. Kaminski et al., Phys. Rev. Lett. **84**, 1788 (2000)
49. A.D. Gromko et al., Phys. Rev. B **68**, 174520 (2003)
50. T. Sato et al., Phys. Rev. Lett. **91**, 157003 (2003)
51. X.J. Zhou et al., Phys. Rev. Lett. **95**, 227002 (2005)
52. W. Meevasana et al., Phys. Rev. Lett. **96**, 157003 (2006)

Chapter 5
Neutron Spectroscopy

Neutron scattering studies with resolution in to the sub-μeV range are now possible by the use of the neutron spin-echo (NSE) technique first introduced by Mezei [1, 2]. In this technique the neutron energy transfer, $\hbar\omega$, can be measured with a resolution several orders of magnitude higher than the energy spread of the incident neutron beam, contrary to the conventional neutron spectrometers.

First generation spin echo spectrometers were very successful in the study of quasi-elastic and only weakly dispersing excitations. However for the measurement of dispersive excitation like phonons, a special focusing technique, proposed by Mezei in 1977, was required. This focusing technique was realized by Golub and Gähler in 1987 [3] and was implemented on the spectrometer TRISP "triple axis spin echo" at the FRM-2. This chapter is concerned with the neutron resonance spin-echo triple-axis spectroscopy (NRSE-TAS). An overview of the basic features of TRISP and the different operating modes are discussed.

5.1 Triple-Axis Spectroscopy

Since the 1950s, triple-axis spectrometry (TAS) has been the method of choice to experimentally determine energy- and momentum-resolved phonon spectra of solids [4]. The technique uses crystal monochromators to define the initial and the final neutron energies (the first and the third axes of the TAS). The second axis (the sample) defines the momentum transfer. Energy and momentum conservation requires:

$$\begin{aligned} \vec{k}_i - \vec{k}_f &= \vec{Q}_o = \vec{q}_o + \vec{G} \\ \frac{\hbar^2}{2m}(k_i^2 - k_f^2) &= \hbar\omega_o \end{aligned} \tag{5.1}$$

P. Aynajian, *Electron–Phonon Interaction in Conventional and Unconventional Superconductors*, Springer Theses, DOI: 10.1007/978-3-642-14968-9_5,

where $\vec{k}_{i(f)}$ is the wave vector of the initial (final) neutron, and $\vec{q}_o$ and $\hbar\omega_o$ are the wave vector and the energy of the elementary excitation. $\vec{G} = \vec{a}^*h + \vec{b}^*k + \vec{c}^*l$ is a reciprocal lattice vector. Rotating the three axes (monochromator, sample, and analyzer) allows one, in principle, to probe nearly any coordinate in energy–momentum space.

The small scattering cross section of neutrons and the limited neutron flux necessitates the use of large beam profiles with finite divergency. The result is a gain in intensity at the expense of energy and momentum resolution. The energy resolution is therefore restricted to the meV range and significantly better resolution are impractical to achieve in almost all circumstances. Although phonon energies can be accurately measured, their linewidths (μeV range in case of electron–phonon interaction) are usually beyond the resolution limit of the TAS.

The TAS resolution function has been treated by Cooper and Nathans and details can be found in [4, 5]. Experimentally the resolution function, which represents a four-dimensional ellipsoid, three dimensions in momentum space and one in energy space, is obtained by performing constant energy or momentum scans at the Bragg peak. Figure 5.1a shows color coded intensity map of the (200) Bragg peak from a perfect $RuMnF_3$ crystal in the $(\vec{Q}, \omega)$ space. The measurement was carried out on TRISP, operated in the TAS mode. The measured signal is a convolution of the spectrometer resolution function and the scattering function of the sample. Since the Bragg peak of a perfect crystal is a delta-function in both $\vec{Q}$ and ω space, the signal displays the instrumental resolution function.

To measure the phonon energy, by performing a constant energy or a Q-scan for example, the resolution ellipsoid is scanned through the phonon dispersion. The observed width depends on the orientation of the resolution ellipsoid with respect to the dispersion surface. In the ideal focusing configuration the long axis of the resolution ellipsoid is parallel to the surface. This results in a narrow phonon peak (TAS focusing condition). In the contrary case, when the long axis is orthogonal to

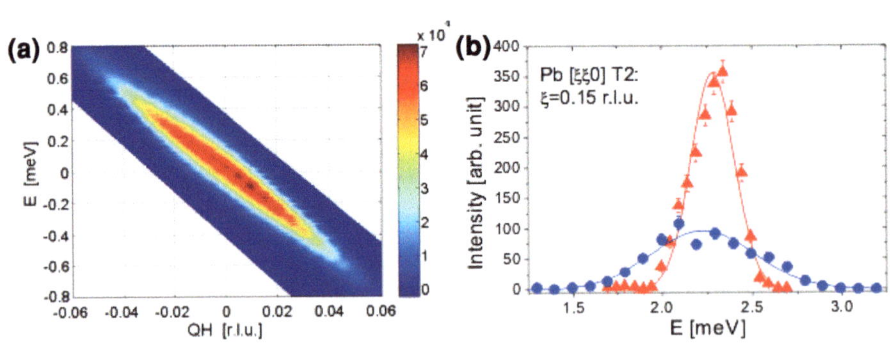

Fig. 5.1 **a** Color coded intensity map of the TAS resolution ellipsoid in the energy–momentum space. The measurements were performed in the TAS mode with fixed $k_f = 1.7$ Å^{-1} (the NRSE components switched off). **b** Focusing and defocusing TAS scan of the (0.15 0.15 0)$T2$ phonon in Pb. The observed width is a convolution of the resolution ellipsoid with the intrinsic linewidth

the surface, the phonon peak will be much wider (TAS defocusing condition). In a typical case, one has no perfect focusing, instead there will be a choice between more focused and less focused configurations (see Fig. 5.1b). Since the TAS energy resolution is in the meV range, deconvoluting the measured peak with the resolution ellipsoid, to obtain an intrinsic linewidths in the μeV range is unrealistic.

5.2 Neutron Spin Echo Spectroscopy

A neutron spin-echo method to improve the energy resolution of TAS by several orders of magnitude without loss of intensity was proposed by Mezei [6], almost 30 years ago. The technique exploits the neutron's intrinsic angular momentum (spin) to access extremely high energy resolution [7]. The underlying physics is based on Larmor precession. It is implemented by inserting spin polarizers and tunable Larmor precession coils into the incident and scattered beams of the triple-axis spectrometer (see Fig. 5.2).

The neutron spin of an initially polarised beam Larmor-precesses in the first coil and reverses its precession in the second coil (where the field is reversed; see Fig. 5.2). The net Larmor precession angle ϕ at the exit of the second coil becomes:

$$\phi = \phi_1 + \phi_2 = \omega_1 \frac{L_o}{v_1} - \omega_2 \frac{L_o}{v_2} \tag{5.2}$$

where $\phi_{1/2} = \omega_{1/2} L_o / v_{1/2}$ is the Larmor precession phase in the first/second arm, $\omega_{1/2} = \gamma_n B_{1/2}$ is the Larmor frequency in the first/second arm with γ_n being the gyromagnetic ratio, L_o is the length of the arm, and $v_{1/2}$ is the initial/final neutron velocity. For elastic scattering ($v_1 = v_2$) and for identical field integrals, i.e. $\omega_1 = \omega_2 = \omega_L$, the net Larmor precession will be zero and the full polarisation

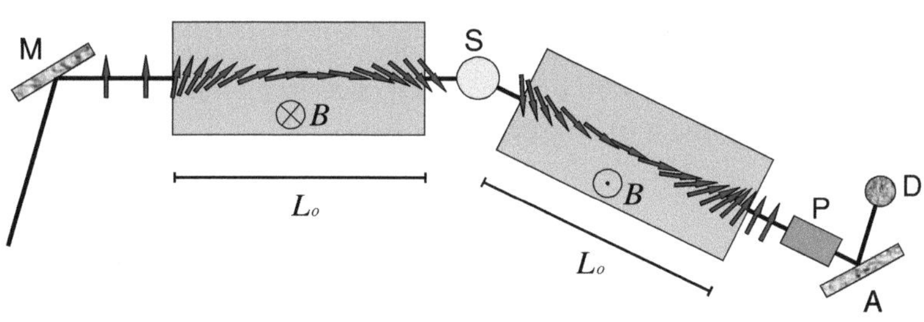

Fig. 5.2 Schematic layout of a classical neutron spin echo setup (*top view*). *M*, *S*, and *A* denote the monochromator, the sample, and the analyzer (the three axes of the TAS instrument). A polarised neutron beam (*arrows*) Larmor precesses in the first coil of constant magnetic field *B* and reverses its precession in the second coil. The final beam polarisation is measured by a supermirror polariser (*P*) and detected at the detector (*D*)

will be recovered at the detector. This is the neutron spin echo. For the case of inelastic scattering, where the neutron loses an energy $\hbar\omega$ and creates a phonon, the total net Larmor precession becomes:

$$\phi = \phi_1 + \phi_2 = \omega\left(\frac{2\hbar\omega_L L_o}{mv_1^3}\right) = \omega\tau \tag{5.3}$$

This expression is obtained by using the equation $\hbar\omega = \frac{m}{2}(v_1^2 - v_2^2)$, where m is the mass of the neutron. Note that ω denotes the phonon energy rather than the Larmor frequency which is denoted as ω_L or $\omega_{1/2}$. The quantity in the parenthesis in Eq. 5.3 has dimensions of time and is known as the spin echo time τ. The polarisation at the exit of the second coil is the cosine of the net Larmor phase ϕ,

$$P_f = P_o\langle\cos(\omega\tau)\rangle = P_o\int R(\vec{Q},\omega)S(Q,\omega)\cos(\omega\tau)\,\mathrm{d}\omega \tag{5.4}$$

where $S(\vec{Q},\omega)$ is the scattering function, which defines the probability of a neutron to be scattered from the sample with energy transfer $\hbar\omega$ and momentum transfer $\vec{Q}$ and $R(\vec{Q},\omega)$ is the TAS transmission function or simply the resolution ellipsoid, which defines the probability of a neutron with energy $\hbar\omega$ and wave vector $\vec{Q}$ to be transmitted through the spectrometer. Since the resolution ellipsoid ($R(\vec{Q},\omega)$) has an energy width in the sub-meV range, whereas the intrinsic phonon linewidth ($S(\vec{Q},\omega)$) is usually in the μeV range, $R(\vec{Q},\omega)$ can be assumed to be constant. Equation 5.4 therefore resembles the time Fourier transform of the spectral function $S(Q, \omega)$. A Lorentzian phonon spectral function, centered at ω, with a half width at half maximum (HWHM) Γ, transforms to an exponential decay of the polarisation:

$$P(\tau) = P_o\exp(-\Gamma\tau) \tag{5.5}$$

Neglecting the instrumental resolution, the full polarisation of the incident beam is recovered at the detector if the measured phonon has an infinite lifetime and the precession fields are properly matched (spin-echo condition). If however the phonon lifetime is finite, the net polarisation decays exponentially, and the phonon linewidth can be determined by systematically measuring the polarisation at various τ values (proportional to the precession fields).

5.3 Neutron Resonance Spin Echo for Dispersive Excitations

Early versions of this technique have enabled the determination of the lifetimes of rotons in superfluid helium [8] and optical phonons near the Brillouin zone boundary in germanium [9]. However, a major technical obstacle has prevented more general applications. For dispersive excitations the energy transfer $\omega(\vec{q})$ depends strongly on the momentum transfer $\vec{q}$. Figure 5.3a shows the phonon dispersion relation with the TAS resolution ellipsoid. The region within the resolution ellipsoid is illuminated by neutrons. Therefore, a finite momentum resolution leads to a spread in energy transfer and thus a degradation of the neutron spin polarisation that masks the effect of the finite phonon lifetime. A spin echo phonon focusing technique introduced by Mezei [6] and Pynn [10] offered a solution to this problem. By tuning the NSE resolution function to the slope of the phonon dispersion, all neutrons scattered from different $\vec{q}, \omega$ (within the ellipsoid) of the dispersion surface of slope $\vec{\nabla}_q \omega_o$, will have a constant phase ϕ. Such a first order focusing is achieved by tilting the boundaries of the precession fields relative to the neutron beam so that neutrons scattered from a linear dispersion with different final energies will spend the same time in the magnetic field regions (see Fig. 5.3b). Since the required tilt angles are of order 10°–50° for acoustic phonons, this cannot be accomplished by the long solenoids used in the early work.

In the neutron resonance spin-echo (NRSE) technique introduced by Golub and Gähler in 1987 [3], the solenoids are replaced by pairs of compact radio-frequency (rf) coils surrounding a field-free region (see Fig. 5.4). This allows larger tilt angles sufficient to match the dispersion relations of most collective excitations in solids [11, 12]. The spin echo condition for dispersive excitations is achieved by tilting the coils in both arms by an angle:

$$\cos\theta_{1,2} = \frac{\vec{k}_{if} \cdot \left(\vec{k}_{if} - \frac{m}{\hbar}\nabla_{\vec{q}}\omega_o\right)}{\left|\vec{k}_{if}\right| \cdot \left|\vec{k}_{if} - \frac{m}{\hbar}\nabla_{\vec{q}}\omega_o\right|} \tag{5.6}$$

and adjusting the coil radio frequencies $\omega_{1,2}$ in both arms to the ratio:

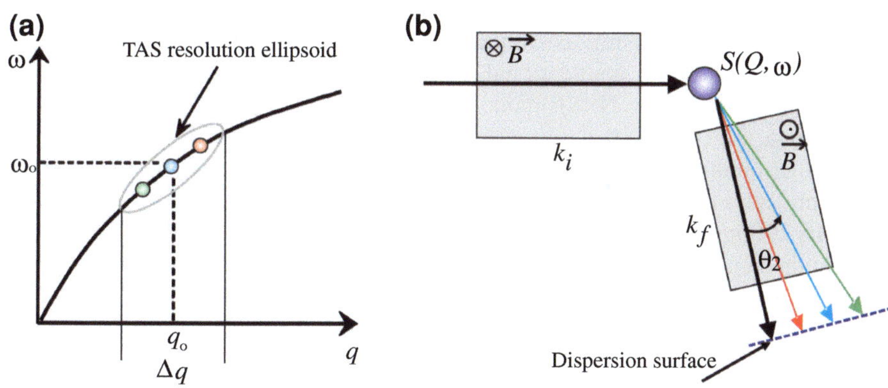

Fig. 5.3 **a** Schematic diagram of an acoustic phonon branch. The ellipse represents the TAS resolution function centered on the phonon (q_o, ω_o). Within the ellipsoid neutrons can scatter from phonons with different energies. **b** The tilting of the field boundaries relative to the scattered neutron beam. By matching the path length in the field to the velocity of the scattered neutrons (determined by the slope of the dispersion, the time of flight of all neutrons in the magnetic field will be the same. The same consideration applies also to the first spin echo arm. This picture was first shown by Pynn [10]

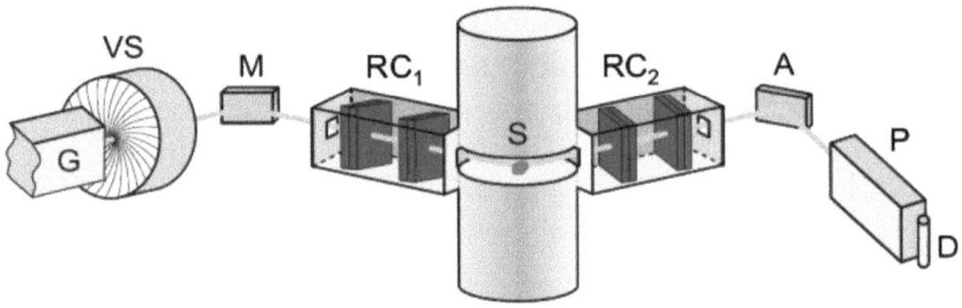

Fig. 5.4 A schematic diagram of the spectrometer TRISP at the FRM-2. *G* denotes the polarizing guide and *P* the transmission polarizer; *M* and *A* are the monochromator and analyzer, as in TAS. *S* is the sample and *D* the detector; *VS* indicates the velocity selector. RC1 and RC2 are the resonance coil pairs which are enclosed by the mu metal shielding boxes. The *line* represents the path of the neutrons through the spectrometer, from left to right in the diagram

$$\frac{\omega_1 L_1}{\omega_2 L_2} = \frac{k_i^3(1 - \frac{m}{\hbar}\vec{k}_i \cdot \vec{\nabla}_q \omega_o / k_i^2)}{k_f^3(1 - \frac{m}{\hbar}\vec{k}_f \cdot \vec{\nabla}_q \omega_o / k_f^2)} \tag{5.7}$$

where $\vec{\nabla}_q \omega_o$ is the slope of the phonon dispersion at ω_o. If these two conditions are satisfied, the spin echo time τ will be the same in both spectrometer arms:

$$\tau = \left(\frac{m}{\hbar}\right)^2 \frac{\omega_{1,2} L_{1,2}}{k_{i,f}^3} \frac{1}{\left(1 - \frac{m}{\hbar}\vec{k}_{i,f} \cdot \vec{\nabla}_q \omega_o / k_{i,f}^2\right)} \tag{5.8}$$

and will be independent of the finite momentum resolution of the TAS spectrometer. This first order focusing technique allows us to measure the linewidth of phonons with μeV resolution within the entire Brillouin zone.

To illustrate this focusing technique and the improved energy resolution of the NRSE compared to the conventional TAS, we mapped out the TAS resolution ellipsoid (Fig. 5.1) in the NRSE mode. Figure 5.5a, b compares the results. In the NRSE mode, we observe the sinusoidal modulation of the intensity. The period of the modulation is inversely proportional to the Larmor frequency and corresponds to the spin echo resolution. The lines of low-and-high intensity exhibit a constant phase. By tilting the coils by θ_1, θ_2, the slope of those lines is tuned to the slope of the phonon dispersion. Figure 5.5c, d compares the phonon linewidth measured in each case. In the NRSE mode, superconductivity induced narrowing is observed as the sample cools below T_c. This effect is masked by the limited resolution in the TAS mode.

5.4 Neutron Resonance Spin Echo for Larmor Diffraction

The neutron Larmor diffraction technique, introduced by Rekveldt [13] allows the measurement of the mosaicity and the $\Delta d/d$ spacing of single crystals independent

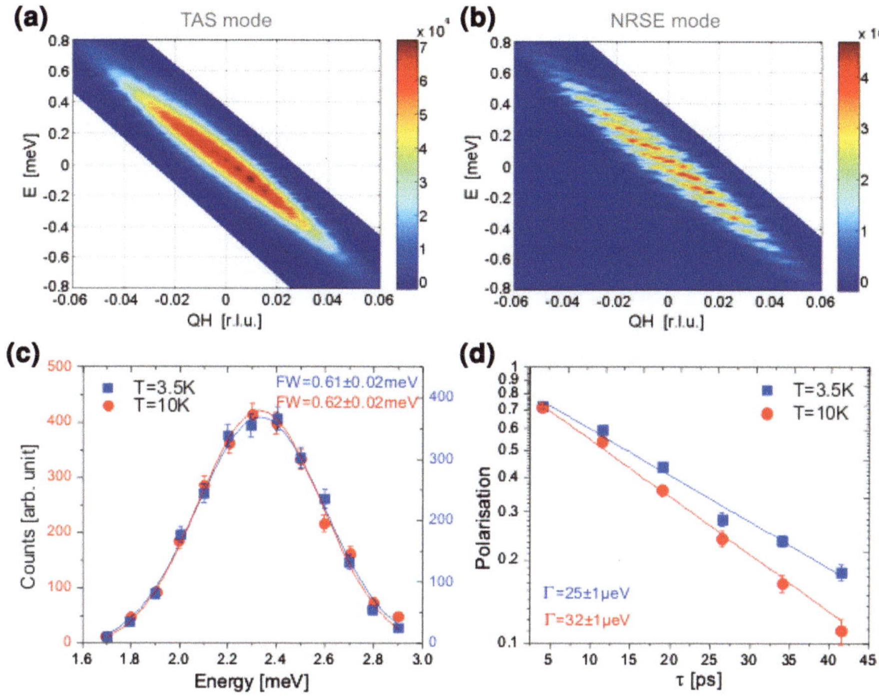

Fig. 5.5 Comparison of the TAS resolution ellipsoid in the **a** TAS and the **b** NRSE modes. **c** Focusing TAS scan of the (0.3 0.3 0)$T1$ phonon in Pb measured below and above the superconducting transition temperature $T_c = 7.2$ K. The observed full widths corresponds to the instrumental resolution. Within the error no change in the linewidths is detected across T_c. **d** The same phonon measured in the NRSE mode. An increase in the full widths of $2\Gamma \approx 14 \pm 2$ μeV is observed above T_c

of each other, and independent of beam divergence and monochromaticity needed in conventional diffraction technique. The technique uses the same NRSE-TAS setup used for linewidths measurement.

Neutron Larmor diffraction for measuring the lattice-spacing spread has been described elsewhere [14]. Here we focus mainly on measuring the mosaic spread of single crystals, which enters as a significant parameter in the resolution function. Figure 5.6 shows a schematic diagram of the Larmor diffraction setup. The rf coils in both arms are tilted parallel to the Bragg planes of the crystal. As in ordinary spin-echo configuration, the sense of Larmor precession is opposite in the two arms (contrary to the measurement of the $\Delta d/d$ spread, where the fields are parallel in both arms). In principle all neutrons that participate in a Bragg reflection from a perfect crystal undergo a zero net Larmor precession in the two arms, independent of the wavelength and direction of the neutrons (solid line in Fig. 5.6). Reflecting on a mosaic block of the crystal with a tilt angle α results in a different path length and thus a different Larmor phase within the first and the

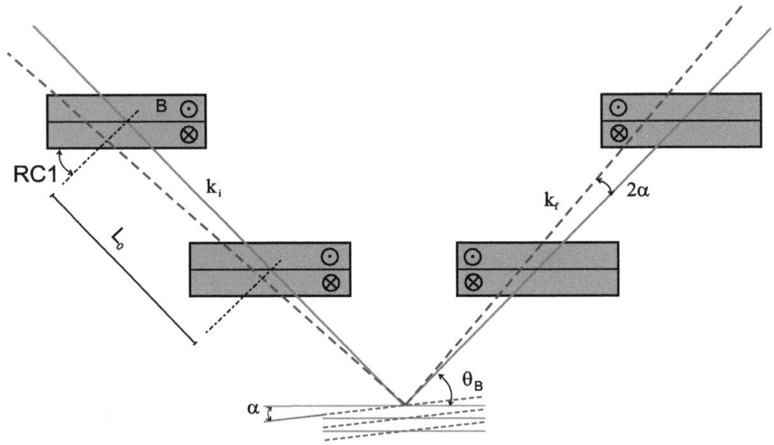

Fig. 5.6 Schematic neutron Larmor diffraction setup for the measurement of the sample mosaic spread. The rf-coils are tilted parallel to the lattice planes of the crystal. The fields in the two spin echo arms are antiparallel. Neutrons scattered from planes parallel to the coils experience a zero net Larmor phase (*solid line*) whereas those scattered from mosaic grains undergo a net Larmor phase (*dashed line*) and depolarise the beam

second arm, as illustrated by the dashed lines in Fig. 5.6. The net Larmor phase, which now depends on the mosaic spread is given by:

$$\begin{aligned} \phi &= \frac{\omega_L L_o}{v}\left(\frac{\sin\theta_B}{\sin(\theta_B+\alpha)} - \frac{\sin\theta_B}{\sin(\theta_B-\alpha)}\right) \\ &= \frac{2\omega_L L_o}{v}\cot(\theta_B)\cdot\alpha \end{aligned} \tag{5.9}$$

Inserting Eq. 5.9 in Eq. 5.4 and assuming a Gaussian mosaic spread for the spectral function and a Gaussian transmission function (rocking scan), the final beam polarization becomes:

$$P_f = P_o \exp\left\{-\left(\frac{2\omega_L L_o}{v}\right)^2 \frac{\cot^2\theta_B}{4\ln 2(1/A3^2_{HH} + 1/\alpha^2_{HH})}\right\} \tag{5.10}$$

where θ_B is the Bragg angle, $A3_{HH}$ is the HWHM of the rocking scan, and α_{HH} is the HWHM mosaic spread of the sample. The mosaicity thus can be extracted by measuring the neutron polarization as a function of the coil frequency.

5.5 NRSE-TAS Spectrometer TRISP at FRM-2

Figure 5.7 shows the NRSE-TAS spectrometer TRISP at the FRM-2 neutron source in Garching, Germany [15]. The compact instrument (length of the spin echo arm is 50 cm) is placed on a thermal beam (excitation energies 1–100 meV)

Fig. 5.7 TRISP at the FRM-2

Fig. 5.8 *Left:* Simple model of a resonance flipper-coil. The static magnetic field $\vec{B}_o$ is vertical to the scattering plane, and the $\vec{B}_{rf}$ rotates in the scattering plane with a frequency ω_{rf}. *Right:* The pair of bootstrap coils in the first spin echo arm. The coils are covered by μ-metal shield. Each flipper-coil can be rotated around a vertical axis by ±50°

at the end of a 10-m long spin-polarizing neutron guide with a critical wavelength $\lambda_c = 0.8$ Å and a transmission of 70% of one spin component. In combination with a horizontally and vertically focusing pyrolytic-graphite monochromator, the polarizing guide clearly outperforms Heusler monochromators. A velocity selector is used to cut out higher-order contamination of the incident beam. A flat pyrolytic-graphite crystal and a supermirror spin-polarizer are used as an analyzer in front of the detector (see Fig. 5.4). A second option of using a horizontally and vertically focusing Heusler analyzer is also possible. A comparison of the different analyzers concerning intensity and resolution is discussed in Chap. 6.

Figure 5.8 shows two pairs of flipper-coils within the first spin echo arm. Each single flipper-coil generates two magnetic fields: a static magnetic field $\vec{B}_o$ along the vertical and a rotating magnetic field $\vec{B}_{rf}$ in the scattering plane rotating with a frequency $\omega_{rf} = \gamma_n \vec{B}_o$ [3, 16, 17]. The coils are shielded with μ-metal and are mounted on rotation stages. The size of the beam windows are 50 × 120 mm^2. The coil frequency ω_{rf} ranges from 50 kHz to 400 kHz[1]. The effective frequency $\omega_{eff} = N \times \omega_{rf}$ where $N = 2$ for the normal mode (2 flipper coils operating in each arm) and $N = 4$ in the bootstrap mode (4 flippers coils operating in each arm). The last bootstrap coil, immediately in front of the detector, can be translated along the beam direction (±15 mm) for performing a spin echo scan.

References

1. F. Mezei, Z. Physik **255**, 146 (1972)
2. F. Mezei, in Neutron Spin Echo, Lecture Notes in Physics, vol. 128 (Springer, 1980)
3. R. Golub et al., Phys. Lett. A **123**, 43 (1987)
4. G. Shirane et al., *Neutron Scattering with a Triple-Axis Spectrometer*. (Cambridge University Press, Cambridge, 1996)
5. M.J. Cooper et al., Acta Cryst. **23**, 357 (1966)
6. F. Mezei, *Neutron Inelastic Scattering*. (IAEA, Vienna, 1978)
7. K. Habicht et al., Phys. Rev. B **69**, 1255 (2004)
8. F. Mezei et al., in *Excitations in Two-Dimensional and Three-Dimensional Quantum Fluids*, vol. 257, ed. by A.F.G. Wyatt, H.J. Lauter (Plenum, New York, 1991)
9. J. Kulda et al., Phys. Rev. B **69**, 045209 (2004)
10. R. Pynn, J. Phys. E **11**, 1133 (1978)
11. T. Keller et al., Phys. Rev. B **241–243**, 101 (1997)
12. T. Keller et al., Physica B **234–236**, 1126 (1997)
13. M.T. Rekveldt et al., Europhys. Lett. **54**, 342 (2001)
14. T. Keller et al., Appl. Phys. A **74**, 127 (2002)
15. T. Keller et al., Appl. Phys. A **74**, 332 (2001)
16. R. Gähler et al., J. Phys. **49**, 1195 (1988)
17. R. Gähler et al., Physica B **229**, 1 (1996)

[1] The magnitude of the maximum frequency is limited mainly by the heat dissipation in the coils. The coils are water and air cooled.

Chapter 6
Experimental Aspects

The neutron resonance spin-echo technique combined with the triple axis spectrometry is a new generation high-resolution spectroscopy optimized for studying the linewidths of elementary excitations. Despite the fact that this technique has been known for more than two decades, TRISP is the only dedicated spectrometer of this kind up to date.[1] Here we review in detail the experimental method and the data analysis. In the first part of this chapter, we concentrate on the experimental details and describe the steps to set the spectrometer and measure the linewidth of a specific phonon. In the second part, we characterize the properties of the different samples used in this work. Of high importance is the NRSE-TAS resolution function. The last part of this chapter is devoted to the instrumental linewidth broadening and resolution correction.

6.1 Experimental Details

Inelastic neutron scattering experiments were carried out at the high-flux NRSE-TAS spectrometer TRISP at the FRM-2 neutron source in Garching, Germany [1]. The spectrometer was operated in the SM = -1, SS = -1, SA = $+1$ scattering senses (SM, SS, SA stand for the scattering sense at the monochromator, sample and analyzer; -1 stands for clockwise and +1 for anticlockwise). With the help of a velocity selector for filtering out high order harmonics, the incident neutron wave vector was varied between $k_i = 2.0 \rightarrow 5.0$ Å^{-1}. The choice of k_i was such that a compromise between resolution and intensity is reached. In general, the highest NRSE resolution is achieved when the phonon dispersion surface sampled by the

[1] There are two additional NRSE-TAS instruments, namely, Flex at HMI and ZETA at ILL. Flex is located at the cold beam and limited to small energy transfer; ZETA is currently in the commissioning phase.

P. Aynajian, *Electron–Phonon Interaction in Conventional and Unconventional Superconductors*, Springer Theses, DOI: 10.1007/978-3-642-14968-9_6,

TAS resolution ellipsoid is minimized. This requires the lowest possible k_i. The PG monochromator, set for the (002) reflection, was focused both vertically and horizontally with curvatures varying around $\rho_V \approx 0.6\ \mathrm{m}^{-1}$ and $\rho_M \approx 0.1\ \mathrm{m}^{-1}$ depending on the choice of k_i. Depending on the required resolution, measurements were performed either with a flat PG analyzer and a supermirror spin polariser (collimation of $\alpha_4 = 60'$) or a vertically and horizontally focused Heussler analyzer. No additional collimations were used.

TRISP can operate in three different modes. (1) The TAS mode (NRSE components switched off), (2) the NRSE mode, and (3) the Larmor diffraction mode (where the precession sense in the two spin echo arms is the same).

To measure the linewidths of a specific phonon (q_o, ω_o), the spectrometer is set in the NRSE mode. The phonon dispersion relation along the high-symmetry directions at low temperature were measured in the TAS mode in the beginning of the experiment. The slope $\vec{\nabla}_q \omega_o$ is obtained from the derivative of a Born–von Karman fit to the phonon dispersion. The coil tilt angles, the frequency ratio, and the minimum τ value are calculated using Eqs. 5.6, 5.7, and 5.8 of the previous chapter. In a TAS focusing configuration, the required coil tilt angles are usually too large for acoustic phonons and are not accessible by the instrument ($|\ \theta_{1,2}\ | < 50°$). For example, the $q = (0.06\ 0.06\ 0)T2$ phonon in Nb has an energy $\hbar\omega = 2.74$ meV and a slope $\vec{\nabla}_q \omega = 16.53$ meV Å. The calculated coil tilt angles are $\theta_1 = 107.09°$ and $\theta_2 = -117.12°$ in the TAS focusing mode which are experimentally inaccessible, whereas $\theta_1 = -26.53°$ and $\theta_2 = 16.86°$ in the TAS defocusing mode are easily accessible. As a result the TAS is driven to a defocusing configuration. In this latter configuration, the segment of the dispersion surface covered by the TAS resolution ellipsoid is minimized, strongly enhancing the momentum resolution.

For the measurement of the linewidth of a particular phonon, the TAS is driven to (Q_o, ω_o) in the defocusing configuration and is kept fixed during the experiment. The coils are tilted by an angle θ_1, θ_2 to match the slope of the dispersion (spin echo phonon focusing). The neutron beam polarisation is measured for several ($\sim$6) τ values by changing the frequencies (ω_1, ω_2) in the flipper-coils, while keeping the ratio (ω_1/ω_2) constant. For each τ, the polarisation P is measured by translating the last flipper coil by a few millimeters (through one period) in the beam direction and extracting the amplitude of the resulting sinusoidal intensity modulation. Figure 6.1a shows the intensity modulation as a function of the coil translation for two different sets of frequencies (ω_1, ω_2) (corresponding to different τ values). The data corresponds to the $q = (0.06\ 0.06\ 0)T2$ phonon in Nb. The polarisation is determined by fitting the data with the function:

$$I(L) = I_o\left(1 + P\cos\frac{2\pi(L - L_o)}{\Delta L}\right) \tag{6.1}$$

where I_o, P, and $2\pi\ L_o/\Delta L$ are the average intensity, the polarisation and the phase of the spin echo signal, and $\Delta L = \hbar k_f / m_n \omega_2$ is the period of the oscillation. Figure 6.1b shows the polarisation extracted from the intensity modulation versus the spin echo time τ. To extract the phonon linewidth, the data are fitted with the

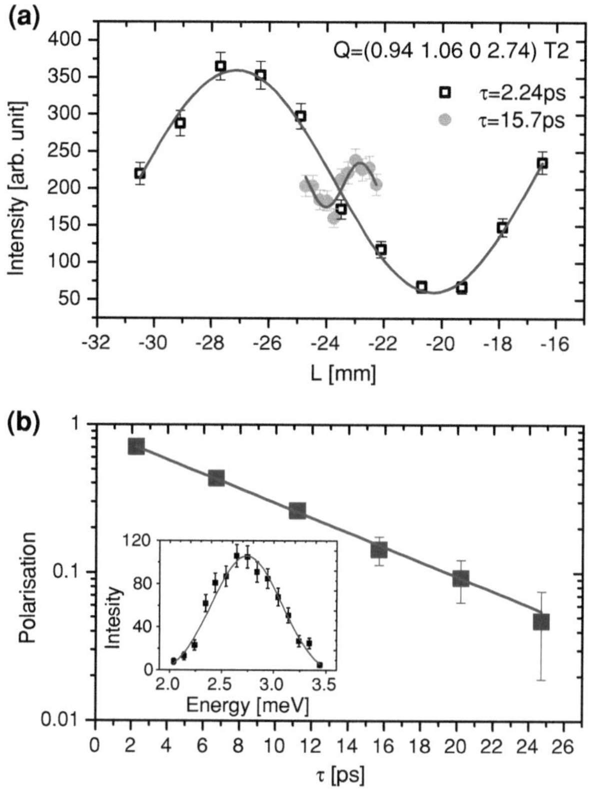

Fig. 6.1 **a** Spin echo intensity modulation of the $Q = (0.94\ 1.06\ 0)T2$ phonon in Nb at different τ corresponding to different coil frequencies. L is the translation of coil-4 with an arbitrary offset (-23.5 mm corresponds to equal arms lengths). **b** NRSE profile of the $Q = (0.94\ 1.06\ 0)$ phonon in Nb measured at $T = 12$ K. The corresponding polarisation values are extracted from the data of **a**. The slope of the line in the semi-logarithmic scale corresponds to the raw phonon linewidth $\Gamma = 76 \pm 4\mu$eV. The *inset* shows the focusing TAS energy scan at $Q = (1.06\ 0.94\ 0)$ performed in the TAS mode

function $P(\tau) = P_o\exp(-\Gamma\tau)$ where Γ is the HWHM of the phonon. On a semi-logarithmic scale, Γ corresponds to the slope of the line.

6.2 Samples

The measurements were performed on high-quality large single crystals. For measurements in the NRSE mode, larger samples (>1 cm^3) are needed. Table 6.1 summarizes the properties of the different samples used. The samples were loaded in a closed cycle cryogen-free ^{4}He refrigerator [2, 3] (see Fig. 6.2) which operates

Table 6.1 Properties of the different samples used in the experiments. The * means that the data were taken from [5]

Sample	T_c [K]	Mosaicity	Size	Scattering plane
$Pb-1$	7.2*	$7.4 \pm 0.2'$	$\varnothing 15 \times 70$ mm^3	(hhl)
$Pb-2$	7.2*	$8.3 \pm 0.2'$	$\varnothing 15 \times 30$ mm^3	$(hk0)$
$Nb-1$	9.31	$4.8 \pm 0.2'$	$\varnothing 30 \times 50$ mm^3	(hhl)
$Nb-2$	9.31*	$6.8 \pm 0.2'$	$\varnothing 30 \times 50$ mm^3	$(hk0)$
$Pb_{0.96}Bi_{0.04}$	7.34	–	$\varnothing 10 \times 10$ mm^3	$(hk0)$
$Pb_{0.88}Bi_{0.12}$	7.66	–	$\varnothing 8 \times 40$ mm^3	$(hk0)$
$NbSe_2-1$ (single)	7.2*	–	$1 \times 4 \times 5$ mm^3	$(hk0)$
$NbSe_2-2$ (mosaic)	7.2*	$\sim 10 \pm 1'$	$5 \times 4 \times 5$ mm^3	$(hk0)$
$La_{1.92}Sr_{0.08}CuO_4$	24*	$11.3 \pm 0.2'$	$\varnothing 8 \times 40$ mm^3	$(hk0)$
$La_{1.86}Sr_{0.14}CuO_4$	36*	$5.9 \pm 0.2'$	$\varnothing 7 \times 35$ mm^3	$(hk0)$
$La_{1.75}Sr_{0.25}CuO_4$	11.3	$7.9 \pm 0.2'$	$\varnothing 6 \times 35$ mm^3	$(hk0)$

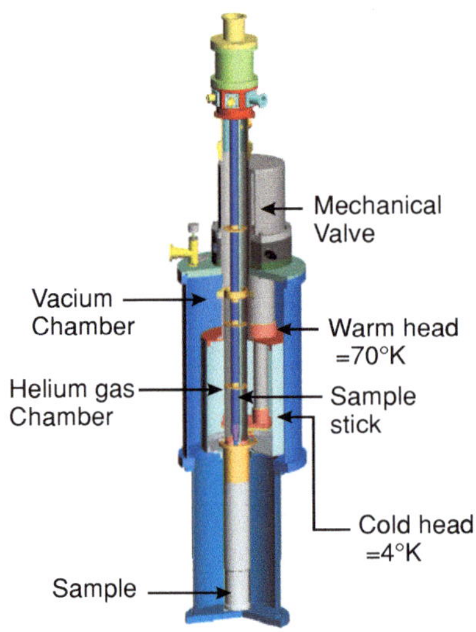

Fig. 6.2 Schematic layout of an FRM-2 type closed cycle cryogen free cryostat with sample stick using pulse tube cooling [4]. Utilization: the sample tube is filled with Helium gas up to atmospheric pressure. Keeping the helium flowing out of the sample tube (to prevent air going in) the sample stick is inserted in carefully. After tightening the clamping ring, the chamber is evacuated and flushed several times with 500 mbar of helium to get air and water out. At the end 300 mbar of helium gas is filled in the chamber. The thermal coupling of the cold head to the sample is via the He gas. Temperature range: 3 to 300 K. Cool down time from 300 to 4 K $\approx$ 4 h

at temperatures between $T = 3$ and 300 K. Measurements at lower temperatures down to 500 mK were performed using a ^{3}He insert.

It is important to identify the properties of the samples, such as the superconducting transition temperature T_c, prior to the experiment. The conventional method to measure the T_c is to carry out temperature dependent resistivity or magnetization

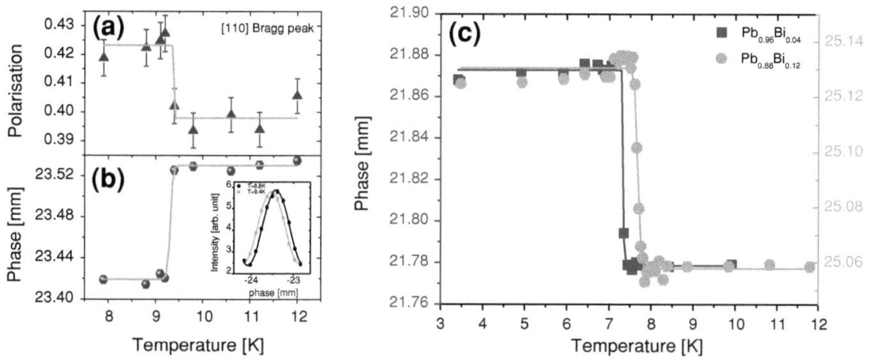

Fig. 6.3 Temperature dependence of the polarisation **a** and the phase **b** of the spin echo signal at T_c under the application of a weak magnetic field. The neutron spin undergos a phase shift as the field is expelled from the interior of the sample below $T_c = 9.32 \pm 0.04\ K$ due to the Meissner effect. *inset*: phase shift of the spin-echo signal. **c** Phase shift of $Pb_{1-x} - Bi_x$ samples with different Bi concentration

measurements. Due to the large sample sizes, this sort of measurements was not possible on a PPMS setup. We have therefore measured the T_c of the large-size samples on TRISP by means of the Meissner effect. The idea was to apply a weak magnetic field on the sample and monitor the shift in the polarisation and the phase of the spin echo signal as the sample cools below T_c and enters the Meissner phase. To generate a weak field, a Helmholtz coil was wound on the outside of the cryostat, symmetrically on each side of the sample, to attain a uniform field at the sample position. Each coil (copper wire $\varnothing = 0.75$ mm) consisted of N=10 loops carrying a current of $I \approx 10$ Amp. The field at the sample site was:

$$B = \left(\frac{4}{5}\right)^{3/2} \frac{\mu_o NI}{R} \approx 9\,\text{Gauss} \tag{6.2}$$

where μ_o is the permeability constant and R is the radius of the cryostat. Due to the Meissner effect, in the superconducting phase, the field is expelled from the interior of the sample, ensuing in a different field integral below and above T_c. Figure 6.3a, b show the polarisation (panel a) and the phase (panel b) of the spin echo signal from the (110) Bragg peak in Nb, and Fig. 6.3c shows the phase shift from the (220) Bragg peak for the $Pb_{1-x}Bi_x$ samples. In all cases a sharp discontinuity ($\sim$0.2 K) in both the phase and the polarisation was observed at T_c. The extracted T_c values are in good agreement with previously reported data [5]. Figure 6.4 shows similar measurements for an overdoped $La_{1.75}Sr_{0.25}CuO_4$ sample. The data show a broad transition ($\sim 8K$) from the normal to the superconducting phase, which can be the result of an inhomogeneity of the Sr atoms in the sample.[2] The T_c of the measured samples are all listed in Table 6.1.

[2] The broad transition can also be an effect of local pairing above the macroscopic T_c.

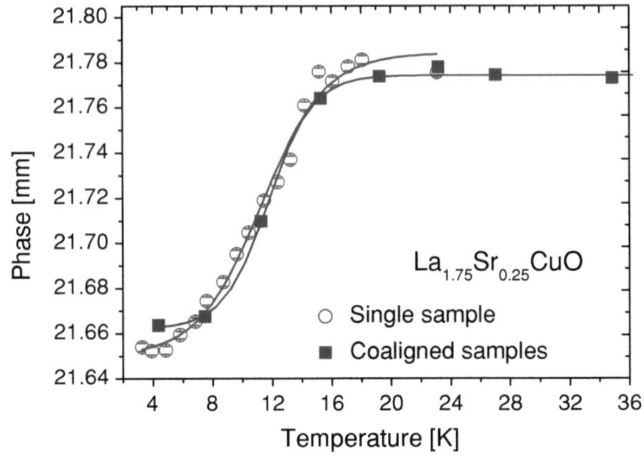

Fig. 6.4 Temperature dependence of the spin echo phase shift in overdoped LSCO. The measurements were carried out on a single crystal (*open circles*) and two co-aligned single crystals (*solid square*). The broad transition is presumably due to the inhomogeneity of the *Sr* atoms in the sample

Another important sample property is its mosaic spread, which is highly significant for the NRSE-TAS resolution function and data analysis. Sample mosaicity is usually measured by gamma diffraction, where the high energy photons can penetrate through the crystal, making it a bulk sensitive technique. Due to the large distance between source-sample and sample-detector of gamma diffractometers ($\sim$5 m), the instrumental resolution is orders of magnitude smaller then the typical mosaic spread of single crystals. This makes it a strong tool to characterize sample quality. The disadvantage of this technique is that only a single spot ($\sim$1 mm^2) on the large-sized crystal can be illuminated, which in some cases can be deceptive.

The neutron Larmor diffraction technique allows measuring the mosaicity of large single crystals independent of beam divergence and monochromaticity. The mosaic spreads of the different samples used in the experiment were measured on TRISP by the Larmor diffraction technique. Details of the technique were given in Chap. 5. Figure 6.5 shows the mosaicity of different selected crystals. A rapid decay in the polarisation indicates a large mosaicity since Larmor diffraction measures the Fourier transform of the mosaic spread. Panel a compares the different LSCO crystals, whereas panel b compares the Pb$-$1 and the Nb$-$1 crystals. The lines are the results of fits to the data. The extracted mosaic spread η are listed in Table 6.1. A close look to the Nb$-$1 data (Fig. 6.5b) shows a bump at large phase values ($\phi_t \sim 5{,}000$) indicating the presence of a small grain within the single crystal. The Gamma diffraction pattern, averaged over many different spots on the same Nb sample, verifies the two-peak feature which quantitatively reflects the Larmor diffraction data (inset of Fig. 6.5b). Both techniques give comparable results within the error bar.

While the samples shown up to now were large single crystals weighing over 10 g, the $NbSe_2$ crystals used in the experiment, were thin platelets (thickness of $\sim$1 mm and a surface of $\sim$10 mm^2). The largest single crystal weighs $\sim$250 mg. In principle, NRSE-TAS experiments for phonon linewidth

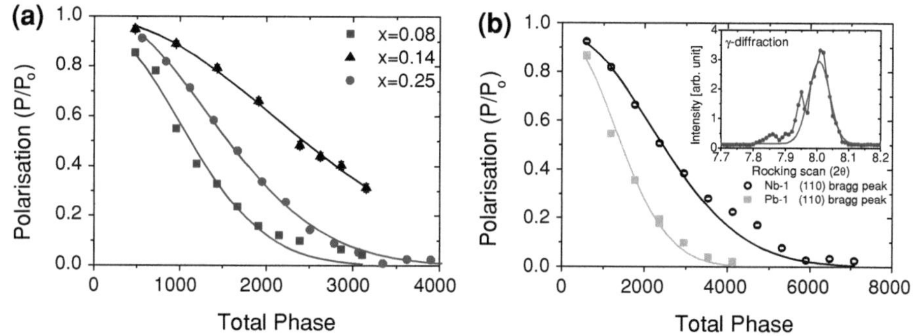

Fig. 6.5 Larmor diffraction from the (200) (x = 0.08, x = 0.14) and the (006) (x = 0.25) Bragg peaks in LSCO (**a**) and from the (110) Bragg peak in the *Nb*−1 and *Pb*−1 (**b**) samples. The polarization is measured as a function of the total Larmor precession phase $\phi_t = 2\omega_L L_o/v$. The *lines* are fit to Eq. 5.10. The *inset* shows γ-diffraction pattern for the *Nb*−1 sample. Both data sets show two-peak feature and when fitted to a single Gaussian (lines in the figure) give similar results

measurement require crystals weighing more than 1 g. A reasonable experiment thus demands to co-align several crystals. Co-aligning single crystals on thin Al plates (similar to what has been done in Ref. [7]) results in a relatively large mosaic spread exceeding one degree [7]. As will be discussed in the next section, such a large mosaic spread can destroy the spin echo signal completely. We have therefore followed a different approach. Figure 6.6a shows an array of five co-aligned $NbSe_2$ single crystals. Each crystal is attached to a tiny three-axis goniometer tower (5 × 5 × 20 mm) and separately aligned on an X-ray diffractometer. The total sample mass is ∼700 mg. The mosaicity of the array was measured by Larmor diffraction and is shown in Fig. 6.6b. The data deviates from the fitted function (which represents the Fourier transform of a single Gaussian) indicating that the mosaic spread in real space does not resemble a single Gaussian peak but rather multiple Gaussians (coming from different crystals). Nevertheless, the different Gaussians, each of an estimated width $\eta \sim 4'$, are misoriented within a range of $\sim 6'$. The effective mosaic spread of the sample array is estimated to be $\sim 10'$. The sample quality is suitable to carry out linewidth measurements, yet, the limited sample mass restricts the measurements to high temperatures and small wave vectors. Table 6.1 summarizes the different properties of the samples used in the experiment.

6.3 The NRSE-TAS Resolution Function

Several nonintrinsic sources lead to degradation of the neutron beam polarisation masking the effect of the finite phonon lifetime. To determine the intrinsic linewidth, the data should be corrected for resolution effects where the combination of

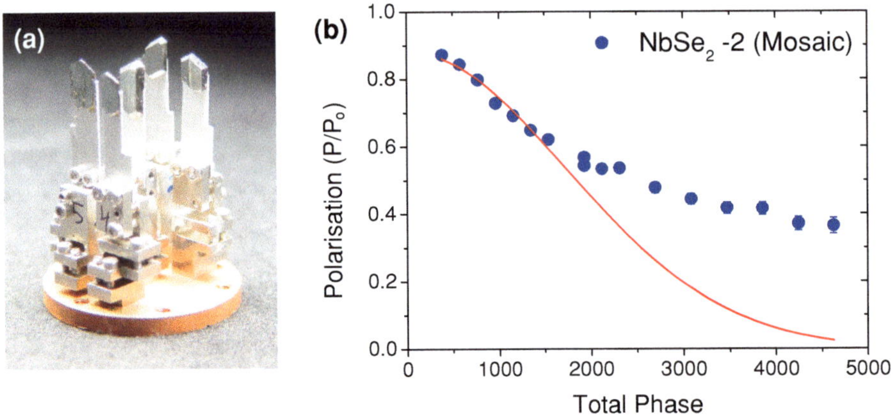

Fig. 6.6 **a** Mosaic crystal array (Mosaic) of five single $NbSe_2$ crystals co aligned each on a single tiny goniometer [6]. The diameter of the copper disc is 4 cm. **b** Larmor diffraction from the (201) Bragg peak of the mosaic samples. The data represents multiple Gaussians coming from different crystals. The mosaic spread of each single crystal is estimated to be 4′ and the effective mosaic spread of the whole sample 10′

both the TAS and the NRSE resolution functions should be taken into account. In spin echo measurements, the TAS resolution ellipsoid defines the "illuminated" four-dimensional (4D) surface of the phonon dispersion around the average value (Q_o, ω_o), whereas the NRSE resolution function gives the artificial (non intrinsic) depolarisation of the neutron beam scattered from this surface. A detailed treatment of the resolution theory of the combined triple-axis (TAS) and neutron resonance spin echo (NRSE) spectroscopy can be found in [8]. In this section we will briefly summarize the effects of the different sources and will use this approach to calculate the resolution function.

The sources of artificial signal depolarisation are instrumental effects and sample properties. The parameters needed for a full resolution calculation are therefore the instrumental parameters (the usual TAS parameters) and the sample dependent parameters, such as the mosaic spread and the curvature of the phonon dispersion surface. Each of these sources will be briefly discussed.

6.3.1 Instrumental Resolution

In the case of inelastic scattering, the spin echo phase $\phi(k_i, k_f) = \omega\tau$ depends on small changes in the initial and final wave vectors, $\delta k^2_{i(f)}$, up to second order. Due to finite beam monochromaticity and divergence, the spin echo phase will have additional terms in $\delta k^2_{i(f)}$, i.e, $\phi(k_i, k_f, \delta k^2_{i(f)})$, which will lead to signal depolarisation. This source, however, does not have a significant effect within the

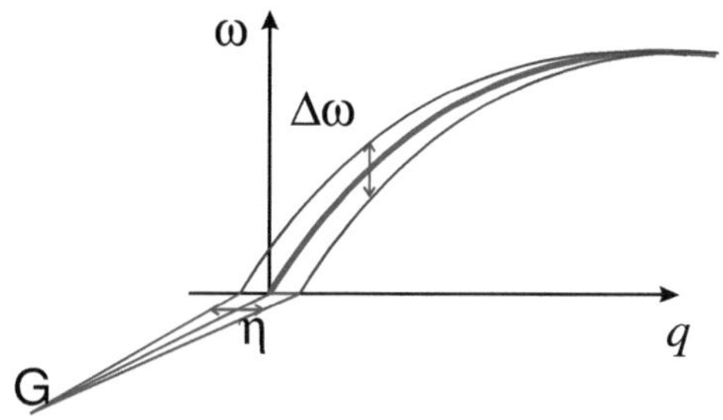

Fig. 6.7 Schematic diagram illustrating the smearing ($\Delta\omega$) of the dispersion surface due to the angular variation in the lattice vector G. The smearing strongly depends on the slope of the dispersion

experimentally accessible range of the spin echo time τ ($\tau < 70$ ps). As an example, calculation of the instrumental resolution for the $(\xi 00)T$ phonon in Pb gives less than 1% drop in the polarisation for $\tau < 70$ ps. This effect is thus negligible.

6.3.2 Sample Mosaic Spread

For transverse acoustic phonon modes, a major limit to the resolution comes from the mosaic spread of the crystal. Due to different mosaic grains in the sample the dispersion surface will have different orientations in q-space. This will smear the dispersion surface in energy and hence will broaden the apparent linewidth (Fig. 6.5). Contrary to the instrumental resolution, mosaicity appears as a first order contribution to the spin echo phase. Assuming a Gaussian mosaic spread η, the beam polarisation will drop by the factor

$$(2\pi\eta)^{1/2}\exp\left[-\frac{1}{2}\tau^2|\nabla_q\omega_o(\mathbf{q}_o)|^2|\mathbf{G}|^2\eta^2\right] \tag{6.3}$$

Due to the term $|\nabla_q\omega_o(\mathbf{q}_o)|^2|\mathbf{G}|^2$, a dispersionless mode will not depolarise the beam and thus will have no effect on the resolution (see Fig. 6.7). For longitudinal branches, where $\mathbf{q}_o$ is parallel to $\mathbf{G}$, the mosaicity will only appear in second order.

6.3.3 Curvature of the Phonon Dispersion Surface

The spin echo phonon focusing condition, discussed in Chap. 5, holds only for linear dispersion. Yet, the dispersion surface within the TAS resolution ellipsoid, has sometimes a considerable curvature. The spin echo phase will then have an additional term

$$-\frac{1}{2}\tau\Delta\mathbf{q}^T\widehat{H}\Delta\mathbf{q} \tag{6.4}$$

where $\widehat{H}$ is the Hessian or the curvature matrix of the dispersion surface at the point $(q_{xo}, q_{yo}, q_{zo}, \omega_o)$ and is defined as

$$H_{ij} = \frac{\partial^2 \omega_\sigma(q_x, q_y, q_z)}{\partial q_i \partial q_j}, \quad i,j = x, y, z. \tag{6.5}$$

where $\omega_\sigma(q_x, q_y, q_z)$ is the energy of the phonon of branch σ, and wave vector (q_x, q_y, q_z).

To facilitate numerical treatment, all these resolution terms were implemented in the MATLAB code SERESCAL by Habicht [9].

6.3.3.1 Phonon Dispersion Surface

Due to the absence of an analytical form of the dispersion surface (except for conventional spin excitations), calculation of the curvature matrix is a complicated task. We have approximated the 4D dispersion surface by the expression:

$$\omega_\sigma(q_x, q_y, q_z) = \sum_i A_i \sin\left(\frac{\pi\sqrt{(aq_x^2 + bq_y^2 + cq_z^2 + dq_xq_y)}}{\Lambda}\right) \tag{6.6}$$

The parameters A_i (with $i \sim 6$ terms) and Λ describe the curvature of the phonon branch along a particular high-symmetry direction and can be obtained from a Born–von-Karman fit to the experimental phonon dispersion along a specific high-symmetry direction. The set of parameters a, b, c, and d, describe the curvature of the dispersion surface along the low-symmetry directions, in and out of the scattering plane. They define the deviation from an isotropic (spherical) dispersion surface to an anisotropic (elliptical) one. These parameters (a, b, c, and d) are assumed to be momentum independent and can be obtained from elliptical fits to a number of low-symmetry phonons along different directions. Here we represent the modeled dispersion surface for each sample.

To model the dispersion for Pb, we have used the experimental data (for low-symmetry directions) of Ref. [10]. Figure 6.8a, b show the modeled phonon dispersion surface in the (hk0) plane (Eq. 6.6). The choice of the A_i, Λ, and the (a, b, c, d) parameters are such that the model precisely describes the dispersion surface in the vicinity of the $(\xi\xi 0)T1$ branch. The experimental data along the $(\xi\xi 0)T1$ (data from TRISP) and the constant energy cuts along low-symmetry directions (data taken from Ref. [10]) are also plotted in order to have a direct comparison with the model. Figure 6.8c shows a 2D cut of Fig. 6.8a at a constant energy $\hbar\omega = 3.3\,\mathrm{meV}$. The model (which represents an ellipse in 2D) accurately describes the curvature of the dispersion in the vicinity of the $(\xi\xi 0)T1$ direction. Within the TAS resolution ellipsoid (represented by the blue circle) the model and the data are in perfect agreement. The experimental dispersion, however, deviates dramatically from the model as one moves towards the $(\xi 00)$ direction. To model the dispersion surface in the vicinity of the $(\xi 00)$ direction, a different set of

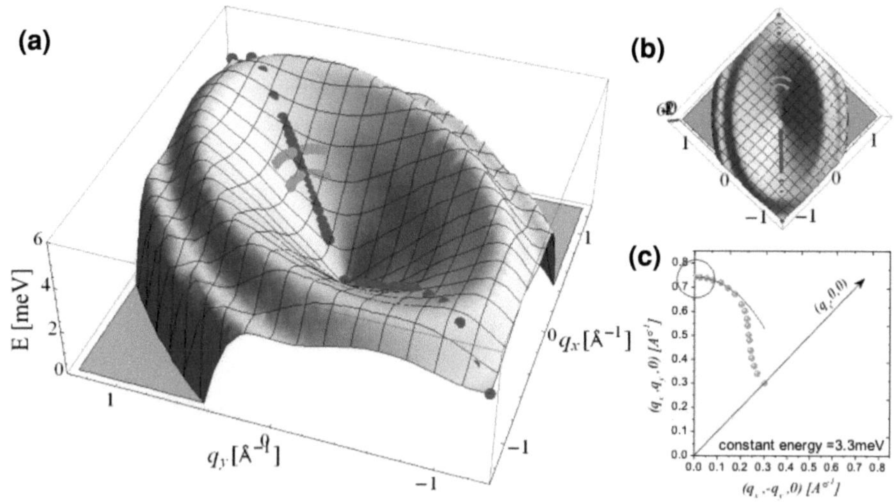

Fig. 6.8 Modeled 3D phonon dispersion surface in the $(hk0)$ plane in Pb from Eq. 6.6. **a** side view, **b** top view. The model precisely describes the curvature in the vicinity of the $(\xi\xi0T1)$ branch. The high symmetry data along the $(\xi\xi0T1)$ direction (data from TRISP) and the off-symmetry data (from Ref. [10]) are plotted for comparison. **c** 2D cut of panel **a** at a constant energy $\hbar\omega = 3.3\,\text{meV}$

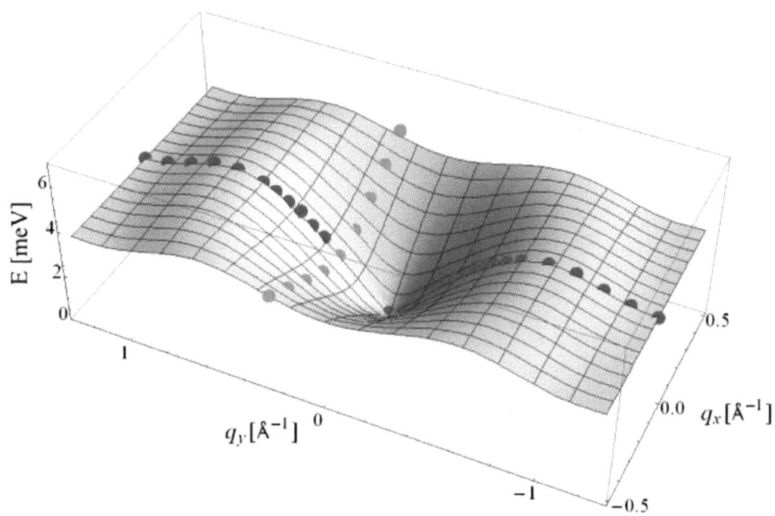

Fig. 6.9 Modeled 3D phonon dispersion surface in the (hk0) plane in Pb describing the dispersion in the vicinity of the $(\xi00T)$ branch. The high symmetry (data from TRISP) and the off-symmetry (from Ref. [10]) data are plotted for comparison

parameters was used. Figure 6.9 shows the modeled dispersion for the $(\xi00)T$ branch in the $(hk0)$ plane. The experimental data are in good agreement with the model.

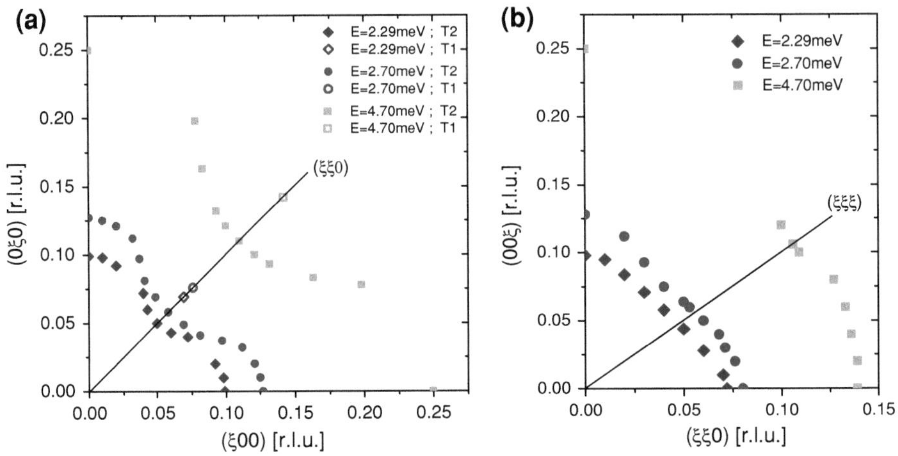

Fig. 6.10 Constant energy cuts of the transverse phonon dispersion in the (**a**) (*hk*0) and (**b**) (*hhl*) planes in Nb at three different energies. The data was taken at $T = 10$ K. The *lines* denote the high-symmetry directions

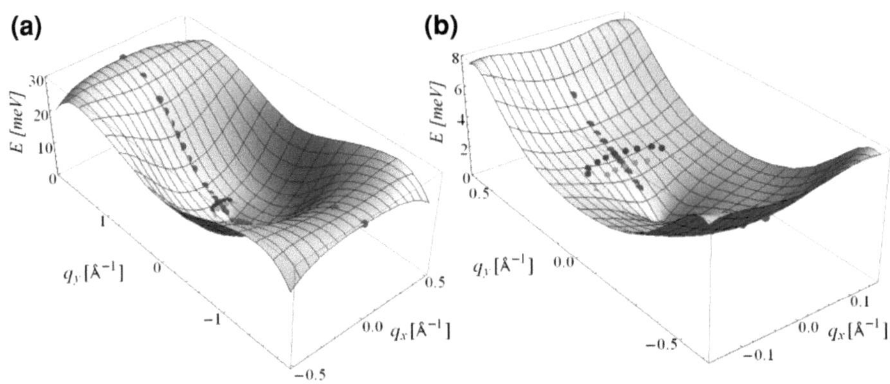

Fig. 6.11 Modeled 3D phonon dispersion surface in the (*hk*0) plane in Nb from Eq. 6.6. **a** Up to Brillouin zone boundary, **b** zoomed in the small *q* region. The model precisely describes the curvature in the vicinity of the ($\xi 00T$) branch. The high symmetry and the off-symmetry data (from TRISP) are plotted for comparison

6.3.3.2 Nb

Off-symmetric phonon dispersion data are lacking for the other investigated materials (Nb, LSCO). To model the dispersion surface in Nb, we have measured the off-symmetry dispersion along several constant energy cuts. The measurements were performed mainly in the small-q region, where the dispersion is strongly curved (see Fig. 6.10). The data are sufficient to assemble a reasonable model for the dispersion. Figure 6.11 show the modeled phonon dispersion surface in Nb in the $(hk0)$ plane. The choice of the A_i, Λ, and the (a, b, c, d) parameters is such that the model describes the dispersion surface in the vicinity of the $(\xi 00)T$ branch. The experimental data are also plotted for comparison.

6.3.3.3 LSCO

Due to the limited neutron beam time and the relatively small sample size, the off-symmetric phonon dispersion was not measured in LSCO. Instead, a different approach was followed to obtain these parameters. The phonon linewidth in LSCO was investigated in detail mainly along the $(\xi 00)\Sigma_4$ branch for $0.05 < \xi < 0.35$. The high symmetry of the (100) direction and the small momentum range covered, allowed us to approximate the off-symmetric curvature of the dispersion surface with two momentum independent parameters (rather than the four a, b, c, d parameters for the general case): one for the in-plane curvature ($(hk0)$) and one for the out-of-plane curvature ($(h0l)$). As we will see in the examples below, for low

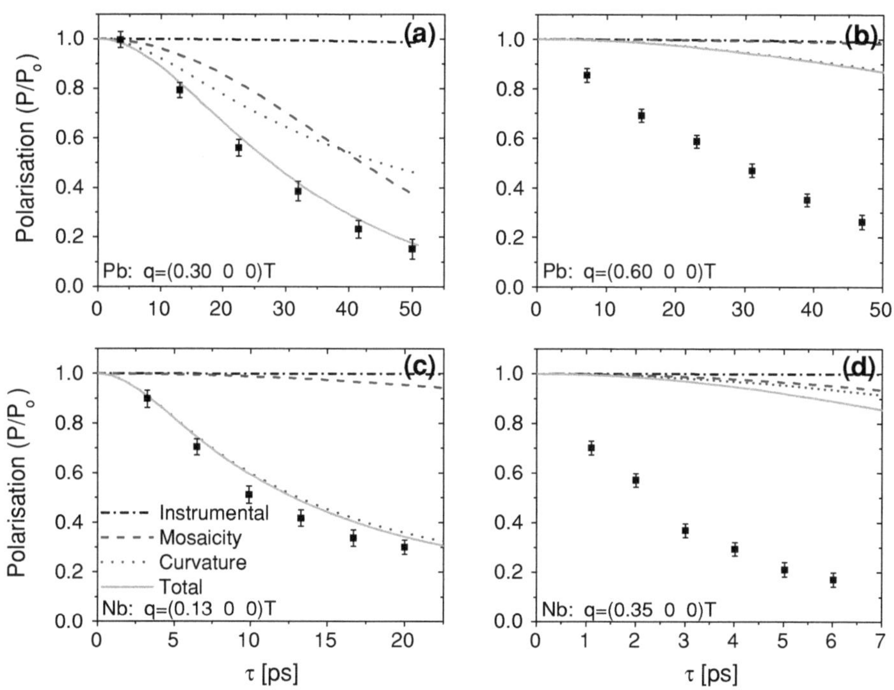

Fig. 6.12 Calculated NRSE-TAS resolution function due to instrumental, sample mosaic, curvature and the total contribution for different selected phonons in (**a**, **b**) Pb and (**c**, **d**) Nb. The experimental data are normalized by a τ independent factor P_o to have a direct comparison with the calculation. For small-q, the data resembles the resolution function. The effect of both the mosaicity and the curvature of the dispersion decrease with q. The resolution terms were calculated using SERESCAL [9]

temperatures the intrinsic phonon linewidth is expected to vanish as $\xi \to 0$. The observed linewidth entirely represents the resolution function. The off-symmetric curvature parameters for LSCO were thus estimated using this assumption.

6.4 Data analysis and Correction

Prior to the correction procedure, the usual TAS parameters were experimentally tested. These parameters, such as beam collimation, initial and final neutron wave vectors, lattice spacing and mosaicity of the monochromator and analyzer, define the dimensions of the TAS resolution ellipsoid. For example, if the ellipsoid is large, the segment of the curved dispersion surface within the ellipsoid will also be large, resulting in a considerable beam depolarisation. In the elastic limit the calculated TAS resolution function was compared with the widths of the measured the Bragg peaks. The experimental (calculated) widths of the (220) Bragg peak in Ge were: longitudinal Q resolution: $\Delta Q_x = 0.030$ (0.031) $Å^{-1}$, transverse Q resolution $\Delta Q_y = 0.018$ (0.017) $Å^{-1}$, and elastic energy resolution $\Delta E = 1.00$ (0.98) meV. The agreement between calculated and measured values gave us confidence in the data analysis and correction. The intrinsic phonon linewidths were obtained by dividing the experimental raw data by the calculated resolution function.

In Fig. 6.12 we compare the calculated NRSE-TAS resolution function with the raw experimental data for different selected phonons in Pb and Nb. The

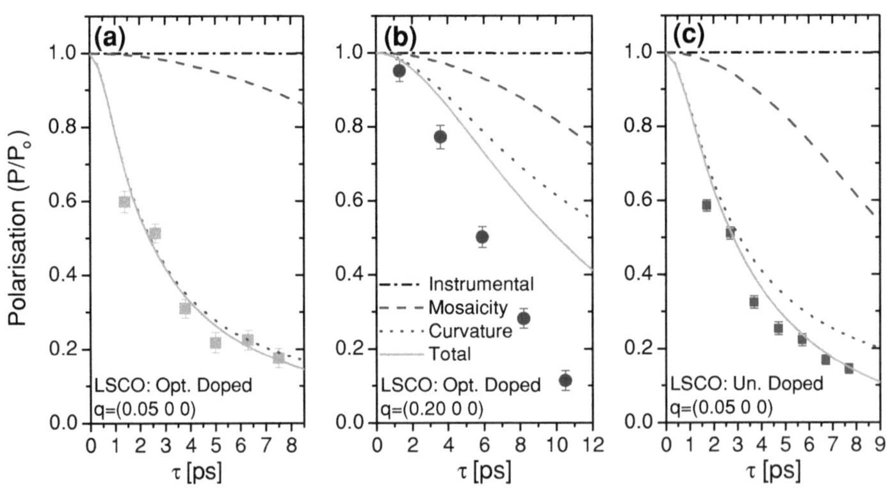

Fig. 6.13 Calculated NRSE-TAS resolution function for optimally doped (**a**, **b**) and underdoped (**c**) LSCO. The experimental data are normalized by a τ independent factor P_o to have a direct comparison with the calculation. The slight drop of the polarization at the minimum τ value is due to the coil transmission

experimental data are normalized with a τ-independent constant (P_o) to have a direct comparison with the calculation. At small-q, the intrinsic linewidth is expected to vanish. The decay of the polarisation can be explained entirely by the mosaicity of the sample and the curvature of the dispersion (panel a, c). Within statistics the intrinsic phonon linewidth is consistent with zero. The effects of both the mosaicity and the curvature strongly decrease with q (panel b, d), whereas the intrinsic phonon linewidth due to the *e–p* interaction in general increases with q. The data are corrected by dividing the experimental polarisation by the calculated resolution function.

Figure 6.13 compares the calculated NRSE-TAS resolution function with the raw experimental data for the optimally doped and underdoped LSCO samples. For $\xi = 0.05$ rlu (panel a), the off-symmetric curvature parameters were adjusted so that the resolution function describes the experimental data. The acoustic phonon dispersion for the underdoped, optimally doped, and overdoped samples, measured on TRISP, did not exhibit any doping dependence within the experimental error bars ($\sim \pm 0.05$ meV), the same curvature parameters were used for

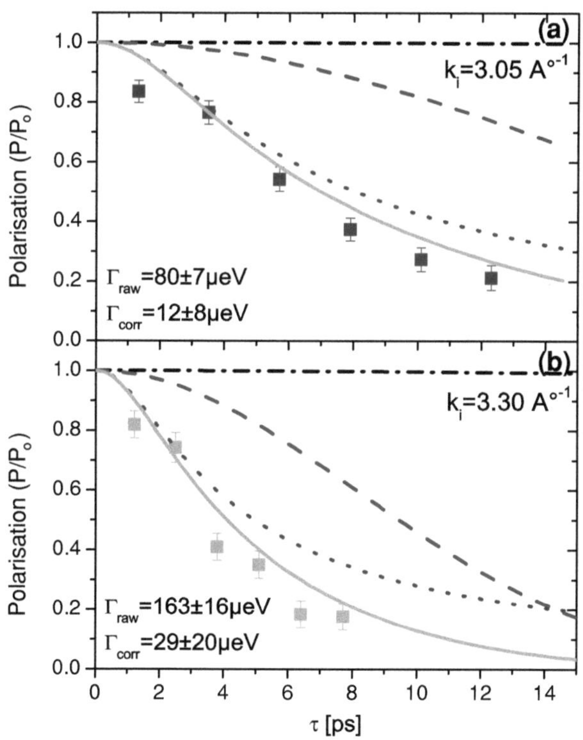

Fig. 6.14 Calculated NRSE-TAS resolution function due for $q = (0.15\ \ 0\ \ 0)\Sigma_4$ phonon in optimally doped LSCO measured under two different spectrometer configurations. The raw and the corrected linewidths are listed in each panel. The notations are the same as the previous figures. The slight drop of the polarization at the minimum τ value is due to the coil transmission

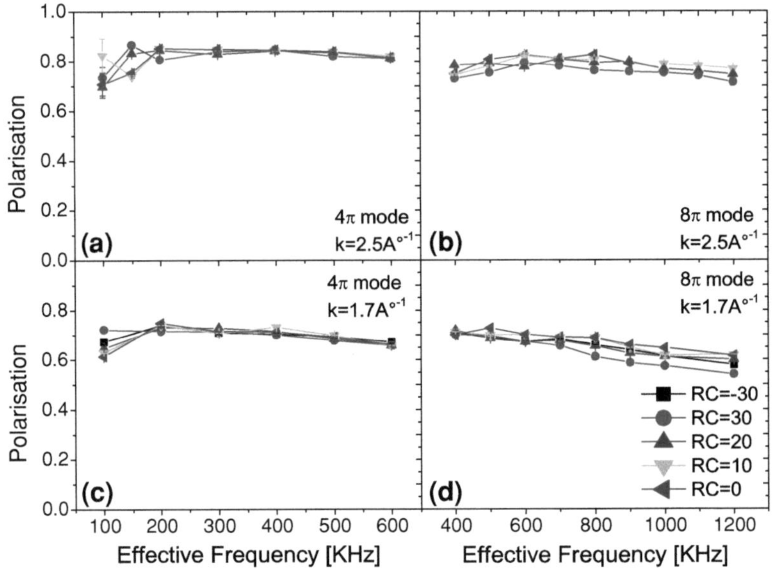

Fig. 6.15 Direct beam calibration for (**a**, **b**) $k = 2.5\ \text{Å}^{-1}$ and (**c**, **d**) $k = 1.7\ \text{Å}^{-1}$. The effective frequency $\omega_{\text{eff}} = N \times \omega_{rf}$ where $N = 2$ for the 4π mode (two flipper coils operating in each arm) and $N = 4$ for the 8π mode (four flippers coils operating in each arm). The slight drop of the polarization at the minimum τ value is due to the coil transmission

all the samples. Panel c shows the $q = 0.05$ rlu phonon for the underdoped sample. The size and mosaicity of this sample is different from that of the optimally doped sample (see Table 6.1). Consequently, the TAS ellipsoid and the instrumental and mosaic contributions to the NRSE resolution function are different for the two samples (panels a and c in Fig. 6.14). Since the curvature correction strongly depends on the shape of the TAS ellipsoid, false curvature parameters result in false resolution function for the two cases. Nevertheless, in both cases the total resolution function agrees well with the experimental data supporting the assumption made earlier. To further verify this point, we have compared the observed and the corrected phonon linewidths of the same phonon measured in two different spectrometer configurations. Figure 6.14 shows the $\xi = 0.15 rlu$ phonon in optimally doped LSCO measured with $k_i = 3.305\ \text{Å}^{-1}$ from the (0 0 6) Bragg peak (panel a) and with $k_i = 3.30\ \text{Å}^{-1}$ from the (1 0 11) Bragg peak (panel b). Despite the fact that the observed linewidths are dramatically different, owing to the difference in the size of the resolution ellipsoid, the corrected data are in good agreement within the error bars, justifying the resolution correction.

In addition to the calculated resolution function, field inhomogeneities and stray fields of the flipper coils also provide a source of signal depolarisation. This is experimentally assessed by measuring the polarisation of the unscattered beam (without sample) as a function of the frequency and the coil tilt angles. Figure 6.15

shows the direct beam polarisation for two different neutron wave vectors. In both cases the polarisation is a rather smooth function of frequency and coil angles and has a negligible effect on the linewidths. All the data were normalized by a direct beam calibration.

References

1. T. Keller et al., App. Phys. A **74**, 332 (2001)
2. Cryostat. www.frm2.tum.de/sampenv.
3. Vericold. www.vericold.de.
4. FRM2. www.frm2.tum.de.
5. J.P. Carbotte, Rev. Mod. Phys. **62**, 1029 (1990)
6. The goniometers were designed by Sible Bayrakci and Benjamin Bruha.
7. V. Hinkov et al., Nature **430**, 650 (2004)
8. K. Habicht et al., J. Appl. Crystallogr. **36**, 1307 (2003)
9. Distributed by Klaus Habicht: habicht@hmi.de.
10. R. Stedman et al., Phys. Rev. **162**, 549 (1967)

Chapter 7
Results and Discussion

This chapter is concerned with the experimental study of the temperature and momentum dependent phonon linewidths in superconductors. The focus is on the electron–phonon contribution to the phonon linewidths. The first part deals with conventional superconductors ranging from simple elements to more exotic two dimensional compounds. The second part is concerned with the high temperature superconductor $La_{2-x}Sr_xCuO_4$ over a wide doping range from underdoped to overdoped samples.

7.1 Conventional Superconductors

Using neutron resonance spin echo spectroscopy, we studied the phonons linewidths in elemental superconductors Nb and Pb (the two elements with the highest superconducting transition temperatures, $T_c = 9.3$ and 7.2 K, respectively). The strategy of the earliest performed experiments on TRISP was to single out the *e–p* contribution to the phonon lifetime by monitoring the evolution of the phonon lineshape through the superconducting transition temperature, T_c. As demonstrated previously by TAS in Nb [1], and Nb_3Sn [2, 3] (where due to the large linewidths and the perfect focusing condition achieved, the intrinsic linewidths were resolved), the *e–p* contribution vanishes if the phonon energy is below the superconducting energy gap, $2\Delta(T)$. Other intrinsic contributions to the phonon linewidth, such as isotope disorder and the anharmonicity of the lattice potential, do not exhibit anomalies at T_c. Temperature dependent measurements of the phonon linewidths in Pb and Nb have shown a superconductivity induced narrowing below T_c.

With a rigorous treatment of the resolution function, required to extract the *e–p* contribution, we studied the momentum dependence of the phonon linewidth in both materials. These investigations revealed a surprising relation between the superconducting gap, extracted from the measurements, and the geometry of the Fermi surface, which also leaves an imprint on the phonon lifetimes. To further

P. Aynajian, *Electron–Phonon Interaction in Conventional and Unconventional Superconductors,* Springer Theses, DOI: 10.1007/978-3-642-14968-9_7,

explore this phenomenon, we have altered the geometry of the Fermi surface in pure Pb by doping it with Bi and investigated its influence on the superconducting gap. Phonon anomalies were also investigated in $2D$ charge density wave superconductor $NbSe_2$.

The experimental results on the different investigated samples are presented in the following sections.

7.1.1 Niobium

Prior to the study of the phonon linewidth, the dispersion of the transverse acoustic branches in Nb was measured along the high symmetry directions. The dispersion is required to calculate the NRSE parameters. The measurements were performed at $T = 3.5$ K and are plotted in Fig. 7.1. The results are in good agreement with previously reported high temperature measurements [4]. The temperature and momentum dependence of the phonon linewidths along these two directions were studied in detail. Figure 7.2a shows a typical spin echo decay profile for a selected transverse acoustic phonon along (ξ00) direction at different temperatures. The data are well described by exponentials, corresponding to Lorentzian phonon spectral functions. Deviations from Lorentzian line shapes were not found within the experimental error (except for the (111) direction which will be discussed in the next section). A focusing TAS scan through the phonon is shown in the inset. This phonon, which has an energy $\hbar\omega < 2\Delta(0)$ is incapable of interacting with the electrons since the decay channel via *e–p* interaction is suppressed due to the stability of Cooper-pairs at low temperatures (See Fig. 7.2b). The intrinsic linewidth at the lowest temperature ($T = 3$ K) is therefore expected to vanish. The nonzero observed linewidth is due to instrumental limitations, which has been

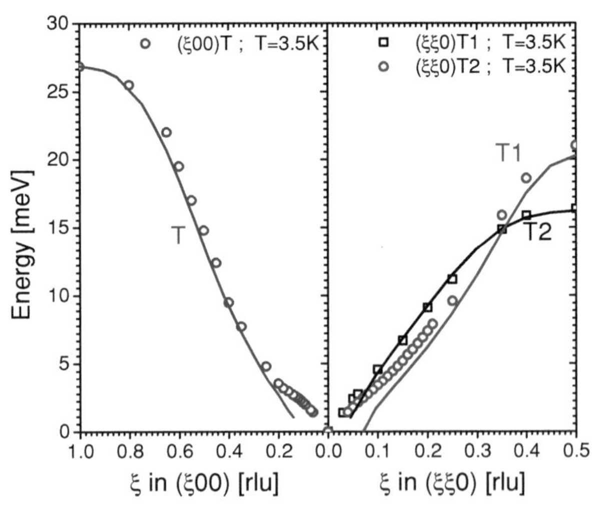

Fig. 7.1 Phonon dispersion of the different acoustic branches in Nb. The data were taken on TRISP at $T = 3.5$ K. *Lines* are the result of lattice dynamical calculations

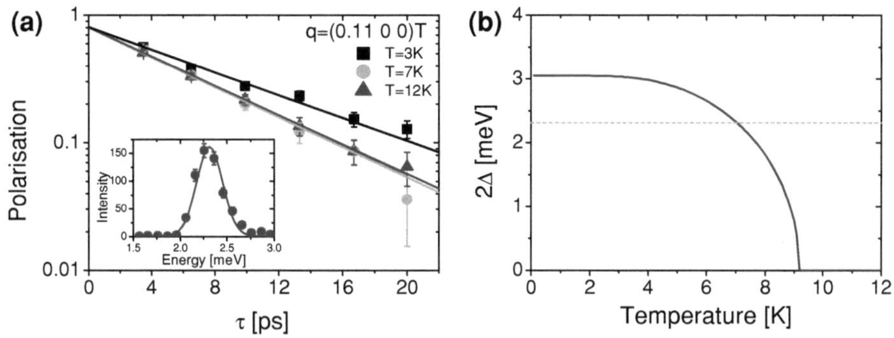

Fig. 7.2 **a** NRSE profile of the $Q = (0.11\ 0\ 0)T$ phonon in Nb at selected temperatures. The *lines* are an exponential fit whose slope in the semi-logarithmic scale corresponds to the raw phonon linewidth . The *inset* shows the focusing TAS energy scan with the NRSE components switched off. **b** The BCS superconducting gap for Nb. The *line* corresponds to the energy of the phonon in (**a**)

quantitatively determined on the basis of the phonon dispersion relations and the mosaic spreads of the single-crystal samples (see previous chapter). These temperature-independent factors have a negligible influence on the temperature dependence of the linewidth. Upon increasing the temperature, the superconducting energy gap is reduced opening an *e–p* decay channel. An increase of the slope (corresponding to an increase of the phonon linewidth) is apparent at $T = 7$ and 12 K. The intrinsic Lorentzian phonon linewidths, Γ, are extracted by fitting the decay profiles to exponentials (lines in Fig. 7.2a) and correcting for the resolution.

Figure 7.3 shows the outcome of a representative set of measurements of the linewidth and energy of several selected transverse acoustic phonons in Nb as a function of temperature. Phonons with energies in the vicinity of the superconducting energy gap are subjected to strong temperature dependence across the transition temperature T_c due to the pile up of the electronic density of states. For phonons with energies $\hbar\omega < 2\Delta(T)$, the *e–p* contribution to the linewidth vanishes as the temperature is lowered below T_c. As a consequence, the intrinsic linewidths Γ abruptly decrease below T_c (Fig. 7.3a–f) and the phonon energies exhibit a corresponding anomaly, as stipulated by the Kramers-Kronig relation (Fig. 7.3f). The effect is no longer present for the $\hbar\omega = 3.30\,\text{meV}$ phonon (Fig. 7.3g) whose energy exceeds the superconducting energy gap. Instead there is some indication of an increase in phonon linewidth at low temperatures. This is expected because of the pileup of electronic states just above the energy gap 2Δ(0). For the more steeply dispersing $(\xi\xi0)T2$ branch, phonons below the gap lie at very small-q, where the electron–phonon interaction is weak (Fig. 7.3h). The superconductivity-induced linewidth narrowing is well described by the BCS excitation spectrum function [5] (lines in Fig. 7.3a–g), and the magnitude and temperature dependence of the energy gap extracted from least-squares fits to corresponding data for various phonons are in good agreement with prior tunneling data and with the BCS expression (Fig. 7.3i).

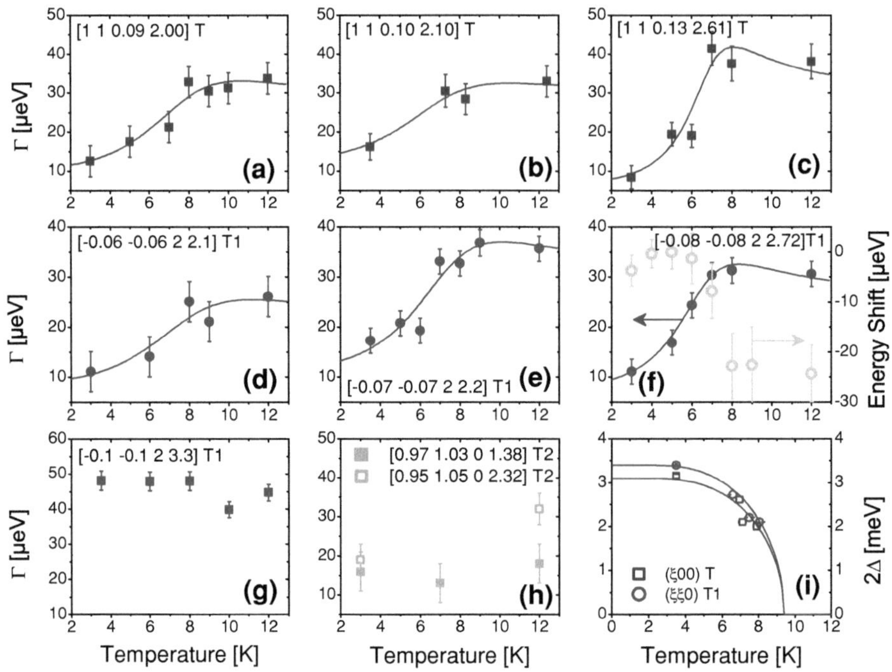

Fig. 7.3 **a–h** Temperature dependence of the intrinsic phonon linewidths Γ for several selected phonons in Nb. The phonon energy shift extracted from the phase shift of the spin echo signals is also shown for the (−0.08, −0.08, 2)$T1$ phonon (panel **f**). The results are Kramers-Kronig consistent. The *lines* are fit to the BCS excitation spectrum function [5]. **i** The BCS expression for the superconducting energy gap in Nb together with the experimental data extracted from panels (**a**)–(**h**). The data at $T = 3.5$ K are extracted from the momentum dependent linewidths of Fig. 7.5

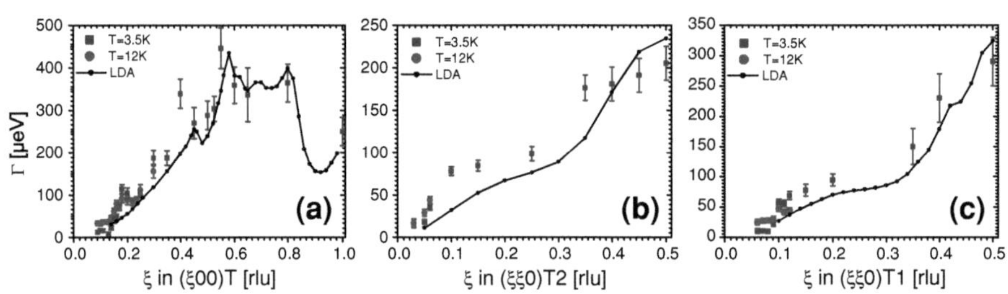

Fig. 7.4 Intrinsic linewidths of transverse acoustic phonons along **a**$q = (\xi 00)T$, **b**$q = (\xi\xi 0)T2$ and **c**$q = (\xi\xi 0)T1$ in Nb at $T = 3.5$ and 12 K. The data were obtained by correcting the measured spin-echo decay rates for all sorts of resolution effects. The connected *solid circles* are the results of ab initio lattice-dynamical calculations [6], as described in the text. *Error bars* indicate the statistical errors

The momentum dependent intrinsic linewidth of the transverse acoustic phonon branches in Nb are shown in Fig. 7.4. The measurements were performed at $T = 3.5$ and 12 K. The overall increase of the linewidth with q is due to the increase of the momentum phase space where electrons can be scattered to. Sharp features are also observed at specific wave vectors in particular along the (ξ00) direction. To help interpret these observations, we compared the results with ab initio density functional perturbation calculation computed on a very fine mesh of q points in reciprocal space [6]. The experimental results are in good overall agreement with the calculations. The resolution of the calculations was sufficient to reproduce the sharp features in q, in particular predicting the subtle structures at large ξ for the (ξ00)T branch. These features are identified as Kohn anomalies arising from nesting of different segments of the Fermi surface.

The data of Fig. 7.4 contain additional exclusive information about the anisotropy of the gap. An acoustic phonon with wave vector $\mathbf{q} = \mathbf{k}_1 - \mathbf{k}_2$ and energy less than the $\mathbf{q}$- dependent gap $2\Delta_{\mathbf{q}} = \Delta_{\mathbf{k}_1} + \Delta_{\mathbf{k}_2}$ cannot decay into electron–hole pairs ($\mathbf{k}_1$ and $\mathbf{k}_2$ are the electron momenta on the Fermi surface). The linewidth of this phonon is not broadened by the e–p interaction below T_c. As its energy, however, exceeds $2\Delta_{\mathbf{q}}$, the linewidth exhibits a discontinuity due to opening of the e–p decay channel. Measurements of this kind, along different crystallographic directions, serve as a comprehensive probe of the superconducting energy gap anisotropy, which can be inferred from a map of these discontinuities. Figure 7.5 displays a blowup of the linewidths and energies of the small-q region in the vicinity of the gap for the (ξ00)T and ($\xi\xi$0)$T1$ branches. The measurements in this region were performed with a fine q-spacing (0.01 rlu), to closely track the discontinuity in the linewidth. At temperatures well below T_c, where the energy gap is saturated (blue squares), sharp jumps reflecting the BCS density of states are apparent in the linewidths when the phonon energy exceeds $2\Delta_{\mathbf{q}}$. The values $2\Delta_{100} = 3.15 \pm 0.05$ meV and $2\Delta_{110} = 3.39 \pm 0.05$ meV extracted from fits to the BCS excitation spectrum function are significantly different, indicating an ~8% anisotropy of the superconducting gap (Note that the error bars quoted include only the statistical error from the fits. The finite q-resolution of the spectrometer translates into a broadening of the observed discontinuities by ~0.12 meV. This systematic error however affects both gap values in the same way and therefore does not influence the gap anisotropy). While both the magnitude and the anisotropy of the gap are consistent with corresponding quantities inferred from prior tunneling data, the phonon measurements were performed in specific crystallographic directions and thus provide an improved basis for a comparison with theoretical calculations.

Similar measurements, performed well above the superconducting transition temperature, reveal the existence of sharp anomalies in the phonon linewidth at the same energies at which the superconducting gap saturates at low temperatures (red circles). These anomalies are Kramers-Kronig consistent with the dispersion anomalies (Fig. 7.5 c, d) and are interpreted as Kohn anomalies resulting from Fermi surface nesting [7]. The data therefore indicate a surprising coincidence between the superconducting gap and Kohn anomalies along both directions.

Presently it is unclear whether the gap energy $2\Delta(0)$ is locked to the Kohn anomalies or if the observed coincidence is accidental. A possible locking

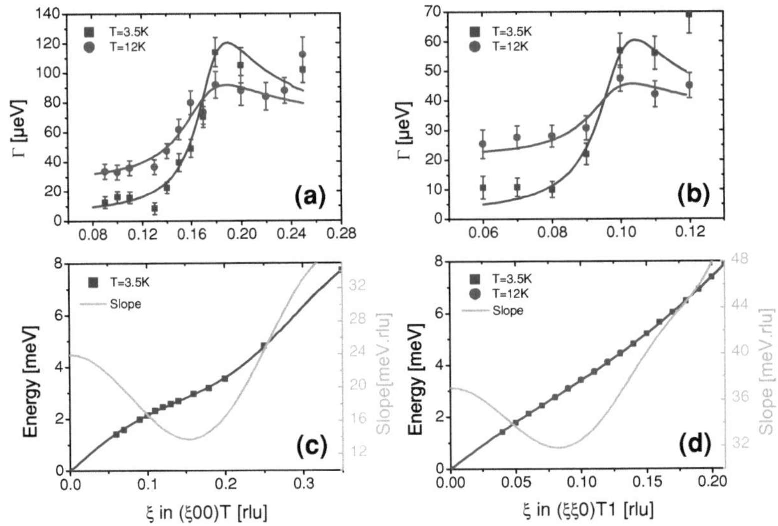

Fig. 7.5 Transverse-acoustic phonon parameters of Nb in the vicinity of the superconducting gap. Phonon linewidths (**a**, **b**) and energies (**c**, **d**) are shown along (1 0 0) and (1 1 0) above (*blue symbols*) and far below (*red symbols*) the superconducting Tc. **a**, **b** The *lines* in the superconducting state are results of fits to the BCS excitation spectrum function. The *lines* in the normal state are guides-to-the-eye. **c**, **d** *Red and blue lines* represent the phonon dispersion (*left scale*), *green lines* represent the numerically computed first derivative (*right scale*)

mechanism was proposed by Scalapino [8]. Such a lock-in effect would limit the $2\Delta(0)$ value to the lowest lying Kohn anomaly. In the following sections we present more experimental data supporting the locking hypothesis.

7.1.2 Lead

Since the early 1960s, the phonon frequencies in lead were investigated with inelastic neutron scattering on the triple axis spectrometer (TAS) [9–11]. Detailed phonon dispersion relations were obtained, but the phonon linewidths were not resolved. Further attempts to increase the TAS resolution in the late 1970s [12], dedicated to the study of superconductivity induced line broadening, were not successful.

The present experiments were dedicated to the study of the *e–p* interaction. The measurements were mostly carried out at temperatures in the vicinity of T_c. Figure 7.6a shows the dispersion of the acoustic phonon branches in Pb along the high symmetry directions. The measurements were performed at $T = 3.5$ and 10 K. Several anomalies are observed in the dispersion, which are in part assigned to Kohn anomalies. Those along the $(\xi\xi0)L$ branch at $\xi = 0.35$ and 0.50 rlu (and also along $(\xi\xi\xi)L$ not presented here) were the first experimentally observed Kohn

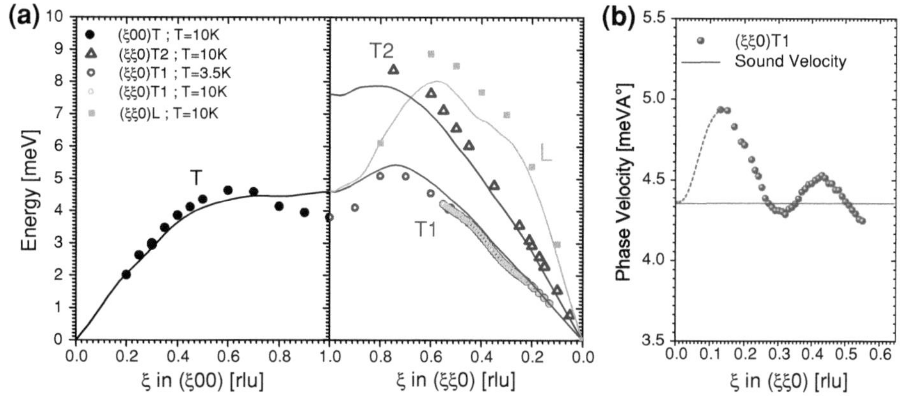

Fig. 7.6 **a** Phonon dispersion of different acoustic branches in Pb. *Lines* are the result of lattice dynamical calculations. **b** Phase velocity of the $(\xi\xi0)T1$ branch. The *horizontal line* corresponds to the sound velocity in Pb at $T = 0$ K [13]

anomalies in metals [10]. They were assigned to diametric Fermi surface nesting. Anomalies at these same wave vectors are also present in the lowest lying $(\xi\xi0)T1$ branch (Fig. 7.6b), which were not quite identified before. Whereas LDA calculations in the framework of ab initio density functional theory presented here [6] predict the Kohn anomalies in the longitudinal branch, do not exhibit any irregularity in the transverse branch around these same wave vectors. Deviations of the LDA from the experimental data are also observed at the zone boundary, where the phonon energies are strongly damped in all the measured branches, in agreement with prior work [10, 11]. These deviations therefore originate from factors not included in the LDA (such as relativistic and spin–orbit effects). Many attempts to explain the observed damping at the zone boundary were made in the past. Spin density wave (SDW) ground state driven by electronic correlations was predicted in Pb [14]. A qualitative theory on the influence of the SDW on the lattice dynamics was successful in explaining the observed dips at the zone boundary [15]. Yet the search for static density waves in Pb was unsuccessful [16].

Recently, ab initio calculations of the phonon dispersion of Pb, within the framework of density functional perturbation theory, with a fully relativistic ultra soft pseudopotential including spin–orbit coupling effects were reported [17]. The dips at the zone boundary and the S-shape anomaly of the T1 branch along the $(\xi\xi0)$, were well reproduced by this method and were shown to be compatible with Kohn anomalies because they appear only for sufficiently small values of the smearing parameter that determines the sharpness of the Fermi surface. The spin–orbit coupling, although on an absolute scale is quite small, turns out to have significant effects in Pb due to the very low phonon energies.

The temperature and momentum dependence of the linewidths of the different phonon branches presented in Fig. 7.6 have been investigated by the spin echo technique. Figure 7.7 displays a typical NRSE decay profile of the transverse acoustic (0.26 0.26 0)$T1$ phonon in Pb at $T = 3.5$ and 6 K. The data are well

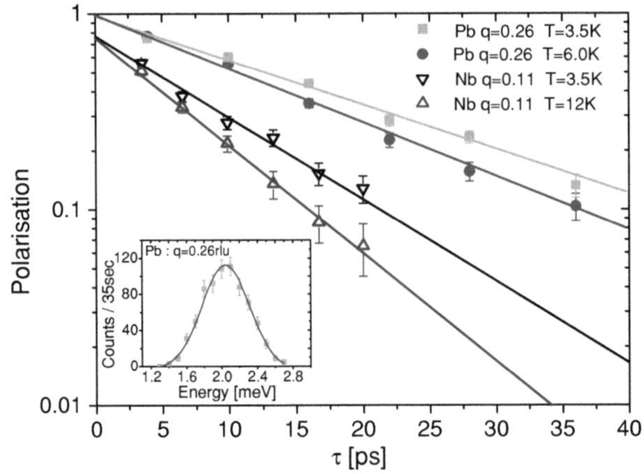

Fig. 7.7 Spin-echo decay profiles of transverse acoustic phonons at $q = (0.26, 0.26, 0)$, phonon energy $E = 2.06$ meV in Pb (*top two curves*) and $q = (0.11, 0, 0)$, $E = 2.32$ meV in Nb (*bottom two curves*) at selected temperatures. The *lines* are the results of fits of exponentials (corresponding to Lorentzian spectral functions) to the data. *Inset* A TAS energy scan through the phonon in Pb. *Error bars* indicate the statistical errors

described by exponentials. A drop of the slope corresponding to an increase of the phonon linewidth is clearly apparent. The corresponding broadening $\Gamma_{\mathrm{Pb}}^{e-p} \approx 5$ μeV is way smaller than $\Gamma_{\mathrm{Nb}}^{e-p} \approx 25$ μeV obtained for Nb. This explains the unsuccessful attempts of observing a superconductivity induced anomalies in Pb by the conventional TAS spectroscopy. The extrinsic factors such as the effect of the mosaicity of the crystal and the curvature of the dispersion relation (where both strongly depend on the slope of the dispersion), which describe the zero temperature linewidth, are larger in Nb due to the steeper phonon dispersion.

Figure 7.8 shows the temperature dependence of the intrinsic linewidth Γ for two $(\xi\xi 0)T1$ phonons with energies below (panel a) and above (panel b) the superconducting energy gap $2\Delta(0)$ (panel c). A decrease of Γ in the superconducting state is evident for the $E = 2.48$ meV phonon. The line represents a fit to the BCS excitation spectrum function [5]. The extracted gap value is plotted in Fig. 7.8c which is in good accord with the magnitude of the superconducting energy gap from tunneling data [18]. The effect is no longer present for the $E = 3.20$ meV phonon whose energy exceeds 2Δ. The sharp structure at $T = 6$ K in panel a and the slight increase in phonon linewidth below T_c in panel b are due to the pileup of the electronic states just above the gap. For the more steeply dispersing $T2$ phonon along $(\xi\xi 0)$ and the T phonon along $(\xi 00)$, the superconductivity-induced phonon linewidth renormalization is much smaller than that of the $T1$ phonon. This is expected on general grounds, because the steep dispersion of these modes imply that the linewidth effects are restricted to wave vectors close to the Brillouin zone center.

Along (111), an accurate determination of Γ was not feasible since the observed NRSE profiles strongly deviate from an exponential decay. Figure 7.9a and b

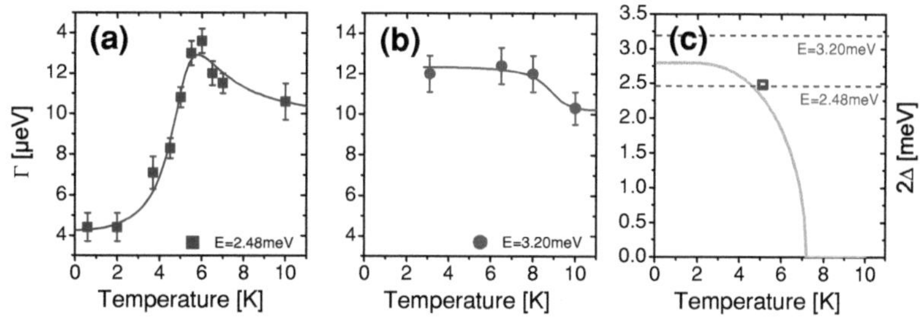

Fig. 7.8 **a**, **b** Temperature dependence of the phonon linewidth for T1-phonons along ($\xi\xi0$). The *lines* in (**a**) is a fit to the BCS excitation spectrum function [5]. The *line* in (**b**) is a guide to the eye. **c** BCS expression of the superconducting energy gap. The *dashed lines* indicate the energy of the phonons in (**a**, **b**). The point is the gap value extracted from the fit to the data in (**a**)

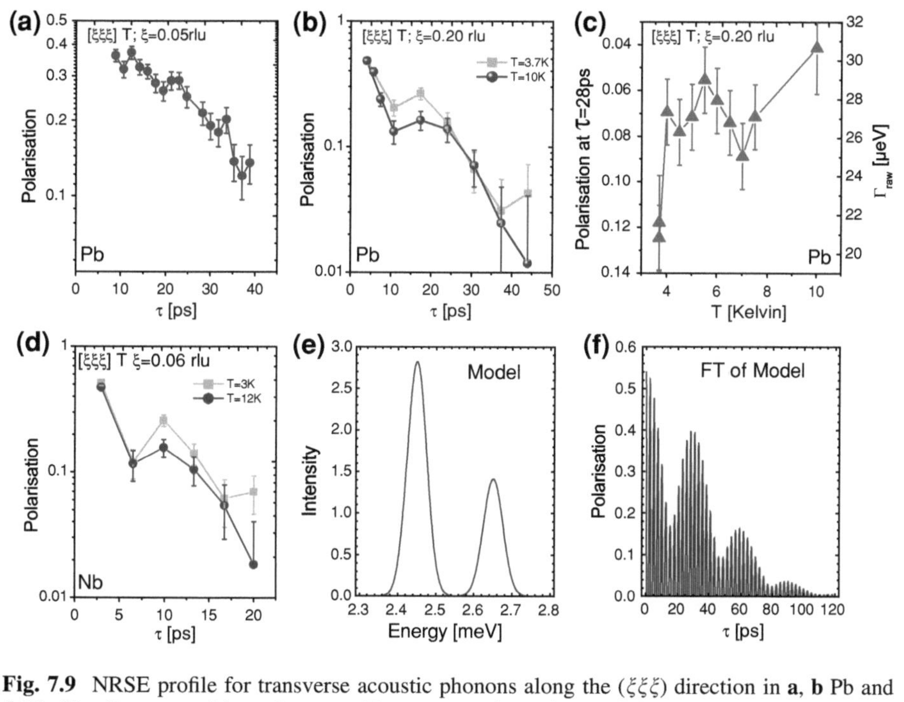

Fig. 7.9 NRSE profile for transverse acoustic phonons along the ($\xi\xi\xi$) direction in **a**, **b** Pb and **d** Nb. The *lines* are guide to the eye. **c** Temperature dependance of the polarisation at $\tau = 28$ ps for the phonon in (**b**). The *right scale* gives the relative estimated raw linewidth. **e** Modeled two Lorentzian peaks corresponding to two phonon groups within the TAS resolution ellipsoid. **f** The Fourier transform of (**e**)

displays the observed NRSE profiles for two different phonons. An oscillation of the polarisation is observed as a function of the spin echo time τ. Even though temperature dependent measurements of the polarisation at a single τ value reveal a superconductivity-induced linewidth reduction of magnitude similar to that of the T1-phonon along ($\xi\xi0$) (Fig. 7.9c), extracting an absolute linewidth is quite complex. Similar deviation from a single exponential decay were also observed for (111) phonons in Nb (Fig. 7.9d). These deviations indicate that $S(Q, \omega)$ does not resemble a single Lorentzian. The observed signal is attributed to the complex dispersion surface of two nearly degenerate transverse phonon branches sampled by the finite q-volume of the spectrometer. Neutrons scattered from both branches with different final energies lead to an oscillation and reduction of the spin-echo signal. To explain this situation, in Fig. 7.9e we present a simple model for $S(Q, \omega)$ with two Lorentzian peaks and in Fig. 7.9f we display its Fourier transform. The result is similar to the observed data. The oscillation and the decay envelope of the polarisation depend on the intensity ratio (which is given by the phonon eigenvectors for the two branches), the widths, and the splitting of the two peaks (in the off-symmetry direction). Due to the complexity of the (111) branch, the momentum dependent measurements were therefore focused mainly along the (100) and (110) directions.

Figure 7.10 shows the momentum dependent intrinsic phonon linewidth along the $(100)T$ and the $(110)T2$ directions at selected temperatures. The data are compared to predictions of LDA calculations [6]. For the low-q region, a fair agreement between the experiment and the LDA is reached. Kohn anomalies along the $(\xi 00)T$ at $\xi \approx 0.35$ rlu and along the $(\xi\xi 0)T2$ branch at $\xi \approx 0.5$–0.6 rlu are observed experimentally and qualitatively reproduced by the theory. Remarkable deviations from the LDA, however, occur at the zone boundary, where phonon broadening is observed in both branches. This huge increase in the linewidths is Kramers-Kronig consistent with the dips observed in the phonon dispersion at the zone boundary (Fig. 7.6). These deviations are believed to be due to spin–orbit coupling, not included in the calculations presented here, which has significant effects on the phonon energies in Pb [17].

The intrinsic wave vector dependent linewidth of the transverse acoustic phonon mode along $(\xi\xi 0)T1$ is shown in Fig. 7.11a. At all temperatures, sharp anomalies in the phonon linewidths are seen at $\xi \sim 0.25$, ~ 0.35, and ~ 0.50 rlu.

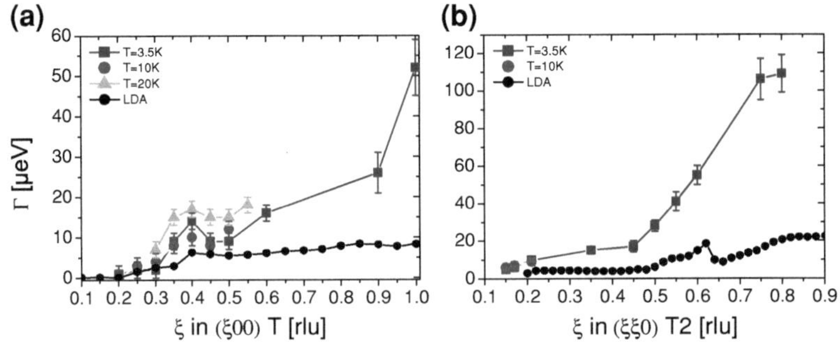

Fig. 7.10 Intrinsic linewidths of acoustic phonons along (**a**) $q = (\xi 00)T$ (**b**) $q = (\xi\xi 0)T2$ in Pb at different temperatures above and below T_c. The data are corrected for resolution effects. The connected *solid circles* are the results of ab initio lattice-dynamical calculations, as described in the text. *Error bars* indicate the statistical errors

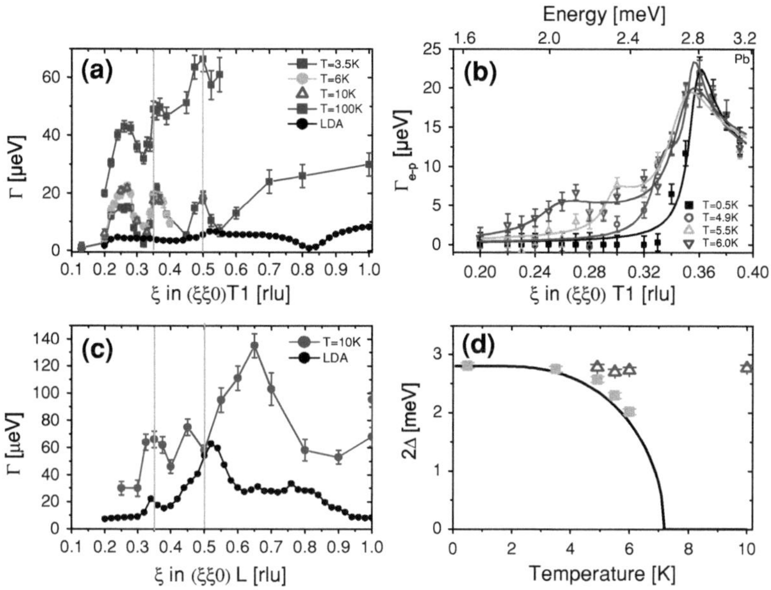

Fig. 7.11 Intrinsic linewidths of acoustic phonons along (**a**) $q = (\xi\xi0)T1$ (**c**) $q = (\xi\xi0)L$ in Pb at different selected temperatures. The data are corrected for resolution effects. The connected *solid circles* are the results of ab initio lattice-dynamical calculations, as described in the text. **b** Contribution of the *e–p* interaction to the linewidth of the transverse acoustic phonon along $q = (\xi\xi0)T1$. The corresponding E is provided by the scale at the top; note the nonlinear E-versus-q relationship (Fig. 7.6). The *lines* are results of least-squares fits to the BCS excitation spectrum function [5]. *Error bars* indicate the statistical errors. **d** The temperature dependence of the superconducting energy gap (*squares*) and Kohn anomaly (*triangles*) extracted from the fits. The *line in the inset* shows the BCS expression for the superconducting gap, which was experimentally confirmed by tunneling spectroscopy [18]. The *error bars* account for systematic errors associated with extracting the peak positions. At $T = 0.5$ K, for instance, fits to two peaks with energies at the end points of the *error bar* are in acceptable agreement with the data

The same features also appear in the phonon dispersion relation (Fig. 7.6): maxima in the phonon linewidth coincide with characteristic S-shaped deviations from the q-linear dispersion, as stipulated by the Kramers-Kronig relation. These phonons with the anomalous wave vectors are intrinsically unstable towards decay into other elementary excitations. In principle, the decay products can be either other phonons (generated, for instance, by anharmonic terms in the lattice potential) or electron–hole pairs (originating from Kohn anomalies). The features at $\xi \sim 0.35$ and 0.50 rlu can be associated with Kohn anomalies, because these wave vectors are known as nesting vectors of the Fermi surface [19]. Indeed, pronounced Kohn anomalies were observed at $\xi \approx 0.35$, 0.45 0.50 and 0.65 rlu in the longitudinal phonon branch [10, 11] (see Fig. 7.11b) and qualitatively reproduced by the LDA assigning their origin to Fermi surface nesting. For the transverse branches, the calculations do not reproduce the sharp features in the linewidths observed by the experiment. Nevertheless, their corresponding appearance in the longitudinal branch indicate that their absence in the transverse branches is not due to poor Fermi surface sampling. The origin of the feature at $\xi \sim 0.25$ rlu is more subtle (Fig. 7.11a), because this wave vector does not match any known spanning vector of the Fermi surface. A possible origin is a three-phonon decay process previously observed in the spectrum of phonons in liquid helium, which are unstable because their phase velocity exceeds the velocity of sound [20, 21]. Indeed, accurate measurements of the dispersion (Fig. 7.6) shows that the phonon phase velocity exceeds the sound velocity around $\xi \sim 0.25$ rlu. This process has not been observed in solids and deserves further investigation. Anharmonic terms in the lattice potential may also contribute to the anomaly.

Now we focus on the influence of superconductivity on the phonon linewidths and explore the validity of the phenomenon observed in Nb. As the superconductor is cooled below T_c, the electron–hole decay channel is closed (and Γ is reduced) below the energy gap 2Δ (T). This effect is observed at low wave vectors ξ in Fig. 7.11a. In particular, $\Gamma \rightarrow 0$ for $T \ll T_c$ around $\xi = 0.32$ (corresponding to a phonon energy of 2.47 meV, below the low-temperature limit of $2\Delta \sim 2.8$ meV known from tunnelling measurements [18]). For lower energies around $\xi \sim 0.25$, however, Γ remains nonzero even at the lowest temperatures, supporting the notion that the linewidth anomaly at this wave vector originates from *p–p* interaction and not from electron–hole pair production. We have removed the contribution of this process for clarity and show only the phonon linewidth $\Gamma_{e\text{–}p}$ directly attributable to the $e-p$ interaction (Fig. 7.11c). As expected, $\Gamma_{e\text{–}p}$ exhibits a maximum due to the pileup of electronic density of states above $2\Delta(T)$, which moves to progressively higher energies upon cooling and closely tracks the energy gap determined in prior tunnelling measurements [18] (Fig. 7.11d). Surprisingly, the superconductivity-induced maximum of $\Gamma_{e\text{–}p}$ merges with the Kohn anomaly as $T \rightarrow 0$. At $T = 0.5$ K, both anomalies are indistinguishable within the measurement error.

In Fig. 7.12a and b we show a blowup of the linewidth of the lowest energy transverse acoustic phonons along the ($\xi\xi 0$) and ($\xi 00$) branches. Kohn anomalies in the normal state ($T = 10$ K) are observed in both directions as sharp increases in the linewidth, and the 2Δ gaps extracted from the low temperature data in the superconducting state coincide with these anomalies along both directions. The nesting wave vectors, that generate the Kohn anomalies, are illustrated in Fig. 7.12c and d by arrows connecting the tubular sheet of the Pb Fermi surface. Once again similar to the observation in Nb, the 2Δ extracted from the low temperature data coincide with the lowest Kohn anomaly along these two directions. While $2\Delta_{110} = 2.80 \pm 0.05$ meV is in agreement with the average gap value for Pb determined by tunneling spectroscopy, $2\Delta_{100} = 3.49 \pm 0.05$ meV exceeds the average by $\approx 20\%$ (top scale in Fig. 7.3) due to locking to the Kohn anomaly along the (100) direction. This surprising lock-in effect, observed in Pb and Nb along different directions, provides a clue to the origin of the superconducting gap anisotropy.

To complement the experimental data, we present LDA calculations [6] of the phonon energies and linewidths in Pb on a grid with a spacing of 0.02 rlu in the

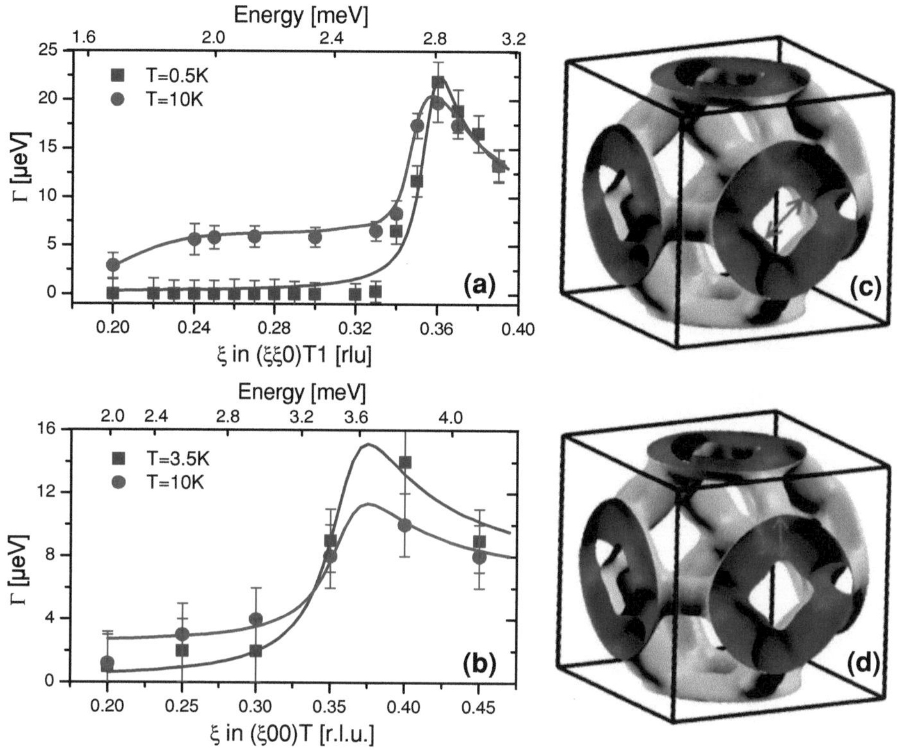

Fig. 7.12 Blow up of the low-q segment of the phonon linewidths in Pb along the $(\xi\xi0)T1$ (**a**) and $(\xi00)T$ (**b**) directions. The *upper axes* give the corresponding energy scale. Note the nonlinear E-versus-q relationship (Fig. 7.6). **c**, **d** The outer sheet of the calculated Fermi surface of Pb, with *arrows* indicating decay channels responsible for the Kohn anomalies in the panels **a** and **b**, respectively

entire $(hk0)$ and (hhl) planes. Fig. 7.13a shows the calculated phonon linewidths along selected directions. For the transverse branches, the calculations do not reproduce the sharp features in the linewidths observed in the experiment above T_c (Fig. 7.12), presumably due to the absence of the spin–orbit coupling [17]. However, Kohn anomalies resulting from nesting of the Fermi surface are clearly seen in the longitudinal branches at momentum points that closely agree with the experimental results, confirming our interpretation of the experimental normal-state features in terms of Kohn anomalies. Figure 7.13b and c show maps of Kohn anomalies extracted from jumps in the numerically computed longitudinal acoustic phonon linewidths in the $(hk0)$ and (hhl) planes (blue symbols). Constant-energy surface of the lowest transverse acoustic phonon branch at a phonon energy corresponding to $2\Delta = 2.8$ meV (solid lines) are also plotted. Remarkably, this surface coincides with a Kohn anomaly within $\pm$ 10% (dashed lines in Fig. 7.13b, c) over the entire Brillouin zone, except for a narrow segment along (100). This implies that lock-in of the momentum-dependent gap to a Kohn anomaly along all

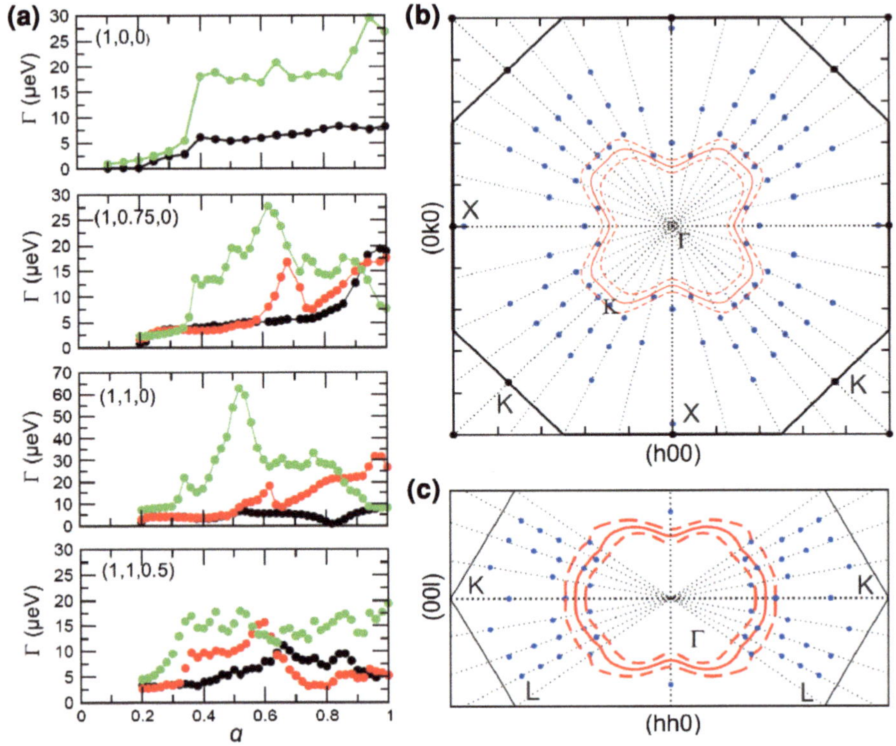

Fig. 7.13 Phonon linewidths and Kohn anomalies in Pb calculated in the framework of ab-initio density functional perturbation theory [6]. **a** Linewidths of longitudinal (*green*) and transverse (*black*, *red*) acoustic phonons along selected directions in the Brillouin zone. Jumps in the linewidths indicate Kohn anomalies due to Fermi surface nesting. **b**, **c** Locus of Kohn anomalies extracted from the jumps in the longitudinal phonon linewidths in the (*hk*0) and (*hhl*) planes (*blue points*). *Solid* (*dashed*) *lines* are the constant-energy surfaces of the lowest-energy transverse acoustic phonon branch at 2.8 meV (2.8 meV ± 10%)

directions in momentum space can generate a gap with an average of 2.8 meV and a slight (±10%) anisotropy over most of the Brillouin zone, in agreement with tunneling data. The lock-in mechanism can also explain the dramatic enhancement of the gap along (100) (Fig. 7.12a), where the Kohn anomaly is located relatively far from the 2.8 meV constant-energy surface. This enhancement has thus far not been reported, presumably because it covers only a very small portion of the Brillouin zone (Fig. 7.13). The superconducting gap anisotropy therefore appears to mirror the locus of Kohn anomalies in momentum space. The lock-in mechanism thus leads to a simple new explanation of the gap anisotropy.

A similarly detailed comparison for Nb is not possible, because at the small wave vectors corresponding to the location of the lowest-lying Kohn anomalies, the numerical values of the phonon frequencies are largely unstable due to strong anharmonic effects.

7.1.3 Lead–Bismuth Alloys

If the superconducting energy gap is controlled by Kohn anomalies and therefore by the geometry of the Fermi surface, varying the latter will correspondingly modify the gap. A possible way to alter the Fermi surface is to apply hydrostatic pressure. Indeed all superconductors exhibit changes in their superconducting transition temperature under pressure. This approach however requires tiny samples, not suitable for inelastic neutron scattering. Doping a material with charge carriers (electrons or holes) changes the dimensions of the Fermi surface and thus systematically displaces the Kohn anomaly to larger or smaller wave vectors. Whether the superconducting gap will follow the Kohn anomaly, supporting the locking effect observed in elemental Nb and Pb, is the aim of the present investigation.

To shed more light on the relation between Kohn anomalies and 2Δ, several crystals of Pb doped with different Bi concentrations were grown [22]. Doping with Bi, which has one electron more than Pb, expands the Fermi surface shifting the nesting to larger wave vectors. 2Δ is therefore expected to increase. Tunneling measurements indeed show a strong dependence of the gap 2Δ and T_c of Pb–Bi alloys on the Bi concentration. Figure 7.14 represents the data taken from [23]. Both quantities dramatically increase with doping. Similar measurements on Pb–Tl alloys (where Tl has one electron less than Pb) show the decrease of the two quantities with doping [23]. Generally speaking, the trend is in the right direction, where in the Bi (electron) doping the FS expands shifting the Kohn anomaly to larger wave vectors, whereas in the Tl (hole) doping, the FS shrinks reducing the nesting wave vectors. We systematically investigated the dispersion and the linewidth of Pb–Bi alloys with different Bi concentrations. The T_c of the high quality samples are plotted (red symbols) in Fig. 7.14 (See the previous chapter for details). The doping and the gap magnitude of the samples are deduced from the T_c measurements. The phase diagram of these alloys indicate that the face centered cubic (FCC) structure persists up to 20% Bi doping [24]. No evidence of contamination from the other phases were observed in the neutron diffraction data for both crystals.

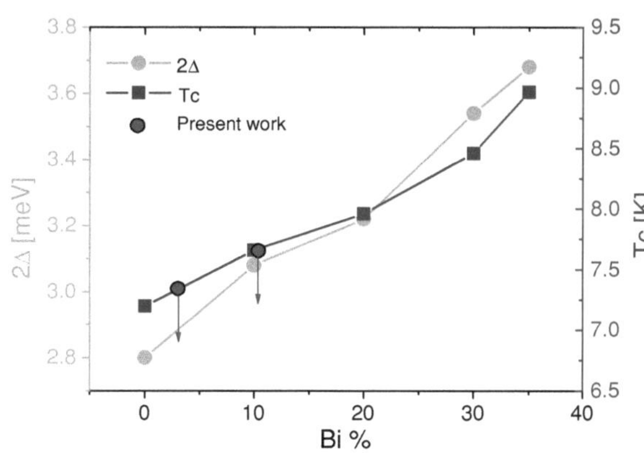

Fig. 7.14 The 2Δ gap and the T_c for the Pb–Bi system. The *red symbols* correspond to the T_c of the samples used in the present work. The data taken from [23]

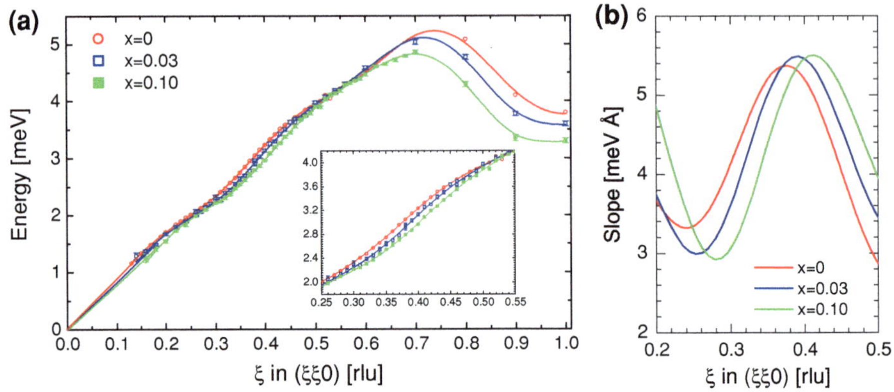

Fig. 7.15 **a** Phonon dispersion of the $(\xi\xi0)T1$ branch in $Pb_{1-x}Bi_x$ alloys with $x = 0$ (pure Pb), 0.03, and 0.10. The *inset* shows a blowup of the anomalous part of the dispersion. **b** The computed first derivative of the Born-von-Karman fit to the dispersion data of (**a**) for the three samples

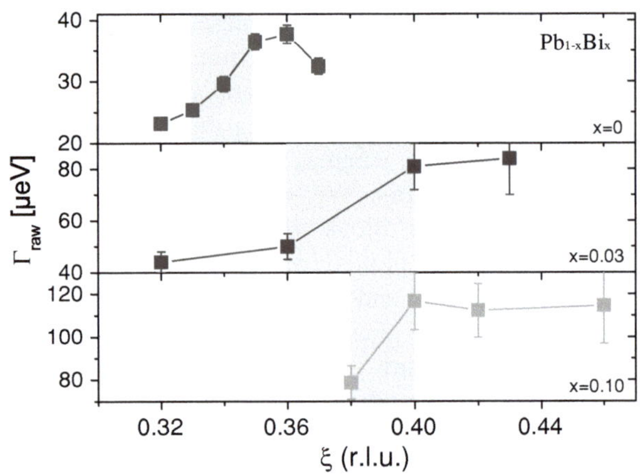

Fig. 7.16 The raw phonon linewidth of the $(\xi\xi0)T1$ branch in $Pb_{1-x}Bi_x$ alloys with $x = 0$ (pure Pb), 0.03, and 0.10. The *shaded* are indicate the widths of the jumps

Figure 7.15a displays the doping dependence of the $(\xi\xi0)T1$ phonon branch in $Pb_{1-x}Bi_x$ samples measured at $T = 10$ K. A clear doping dependence is observed within the ξ-range of 0.35–0.45 rlu. In pure Pb, the $\xi = 0.35$ rlu corresponds to the nesting wave vector where the Kohn anomaly is observed (Fig. 7.12a). Upon progressive doping, we observe the systematic shift of the Kohn anomaly to larger wave vectors (see inset). This shift is clearly seen in the numerically computed first derivative of the Born-von-Karman fit to the dispersion data (Fig. 7.15b). The curves are shifted from the pure Pb by 0.014 rlu for $Pb_{0.97}Bi_{0.03}$ and 0.038 rlu for $Pb_{0.9}Bi_{0.1}$.

The momentum dependent phonon linewidth of the same $(\xi\xi0)$ branch is shown in Fig. 7.16 for the three samples. The vertical shift of the data with doping is most likely due to scattering from disorder (Due to the Bi nuclei which have similar effect as the isotope effect). The increase of the jump magnitude with doping is due to the increase of the electronic density of states at the Fermi surface. Even though dense momentum measurements were not performed (due to the relatively small sample volume and the limited neutron beam time), the presented data are consistent with the shifted anomaly. For pure Pb (panel a), the peak in the linewidth, corresponding to the Kohn anomaly, lies at $\xi = 0.355$ rlu (the value is obtained from a fit to a BCS like excitation spectrum see Fig. 7.12). While the density of the measured points does not allow us to precisely determine the Kohn wave vector for the Pb–Bi alloys, those wave vectors can be estimated from the shift of the slope observed in Fig. 7.15b with respect to the pure Pb sample. The resultant Kohn anomalies lie at $\xi = 0.37$ rlu for $Pb_{0.97}Bi_{0.03}$ and $\xi = 0.39$ rlu for $Pb_{0.9}Bi_{0.1}$. Both values lie within the shaded area of Fig. 7.16 Interestingly, the phonon energies corresponding to the Kohn wave vectors in both cases, 2.84 and 3.0 meV for $Pb_{0.97}Bi_{0.03}$ and $Pb_{0.9}Bi_{0.1}$ respectively, closely trace the superconducting gap 2Δ of Fig. 7.14. This coincident momentum dependence of 2Δ and Kohn anomalies in pure elements and alloys with different crystal structures (BCC and FCC), phonon spectrums, Fermi surfaces, and gap values, therefore is quite unlikely to be accidental, suggesting that the superconducting gap at low temperatures is controlled by the location of Kohn anomalies in the lowest energy acoustic phonon spectrum.

Apart from this phenomenon, the appreciable softening of the zone boundary phonons with Bi doping follows the same trend as the Kohn anomaly observed around 0.35–0.45 rlu. Both become pronounced with doping. Those dips at the zone boundary were also interpreted as Kohn anomalies since they depend strongly on the sharpness of the Fermi surface [17].

7.1.4 Transition Metal Dichalcogenide $2H-NbSe_2$

To test the generality of the phenomenon observed in the elements Nb and Pb and the alloy Pb–Bi, we extended the investigation to more exotic superconductors. Transition metal dichalcogenides (TMD) possess a weakly coupled layered structure. The two-dimensional nature of these materials strongly favors the occurrence of electronic instabilities driven by Fermi surface nesting, which lead to the formation of charge density waves (CDW). Among the different TMDs, $2H-NbSe_2$ has attracted particular attention because of the coexistence of the CDW and superconducting phases. The nature and the competition between these two ground states remains a challenging issue [25].

Conventional CDW transitions are understood in terms of Fermi-surface nesting, leading to the formation of a CDW gap [26]. The CDW and superconducting gaps can be described by the value $\Delta(0) = 1.76 k_B T^*$, where T^* denotes T_c for superconductivity and T_{CDW} for the CDW [26]. It is believed that if these two ground states coexist, they should occur on different parts of Fermi surface. Recent

ARPES experiments on $NbSe_2$ indicated a highly anisotropic superconducting gap which varies among, and even within the different Fermi surface sheets. Surprisingly, the largest superconducting gaps were found to occur at points on the Fermi surface that are directly connected by the CDW ordering vector, i.e. nesting wave vector, in direct contrast to the established view that the two ground states only compete [27, 28].

The phonon spectrum of 2H–$NbSe_2$ has been studied in the past [29]. Detailed measurements of the longitudinal Σ_1 branch revealed a two dimensional Kohn anomaly near half way through the Brillouin zone leading to a CDW. The lower transverse branches were however not carefully studied. We have therefore investigated the dispersion and linewidths of the lowest energy acoustic phonon branch in 2H – $NbSe_2$ single crystals [30]. Detailed measurements with a q – spacing of 0.01 rlu were performed in the vicinity of the expected superconducting gap. Figure 7.17 shows the dispersion of the transverse acoustic (ξ 0 0)Σ_3 branch carried out on a single crystal (SC) weighing $\approx$200 mg at $T = 300$ K. The blue symbols indicate the acoustic branch, whereas the red symbols indicate the optical branch. A kink in the dispersion is observed at $\xi \approx 0.11$ rlu (the lines indicate the change in the slope of the dispersion which intersect around $\xi \approx 0.11$ rlu). Surprisingly, this kink lies at an energy corresponding to the largest superconducting energy gap [28]. Figure 7.18a displays the computed first derivative of the dispersion, which clearly shows an anomaly around $\xi = 0.1$ rlu. A similar kink is observed in the resolution limited phonon widths (FWHM extracted from the TAS energy scans) (Fig. 7.18b).

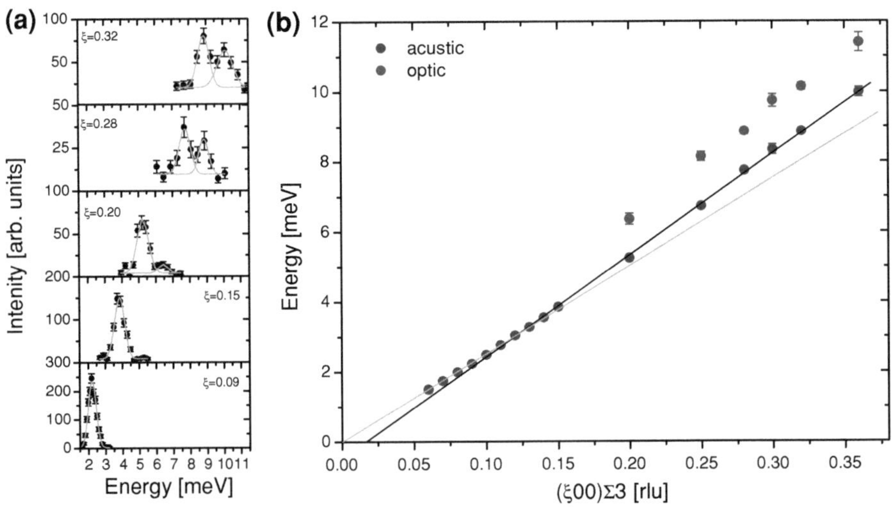

Fig. 7.17 **a** Constant energy scans of the (ξ 0 0)Σ_3 phonon branch in $NbSe_2$. The measurements were performed with fixed $k_f = 2.5\text{Å}^{-1}$. For $\xi > 0.15$ rlu two phonon groups are observed. **b** The dispersion of the (ξ 0 0)Σ_3 branch at $T = 300$ K. The *blue symbols* represent the acoustic branch whereas the *red symbols* represent the optical branch which approaches the acoustic branch for $\xi > 0.2$ rlu. The *lines* are guide to the eye

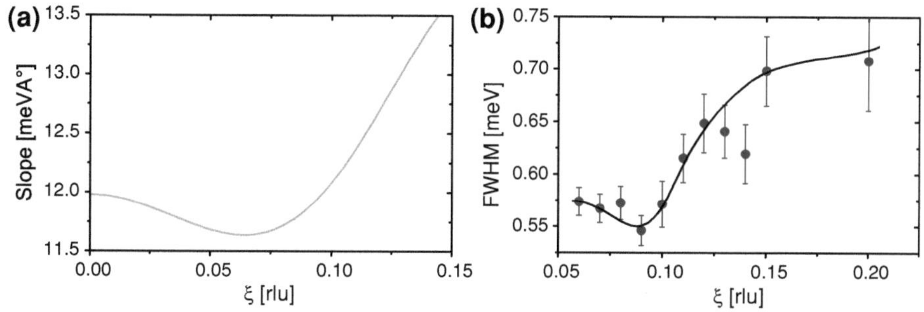

Fig. 7.18 **a** The computed first derivative of the acoustic phonon dispersion (Fig. 7.17). **b** The full width at half maximum (FWHM) of the energy scans as a function of ξ. The width (instrumental resolution) is highly sensitive to the slope of the dispersion. The *line* is guide to the eye

The observed FWHM displays the instrumental TAS resolution which is strongly sensitive to the slope of the dispersion.

A kink in the phonon dispersion can either be caused by the screening electrons (Kohn anomaly) or by the interatomic forces. For layered systems with weak inter-layer forces, a convex curvature is expected for the transverse acoustic mode with out-of-plane polarisation ($\Sigma TA_{\perp}$). Due to the crystal symmetry, the slope of this branch should be equal to that of the transverse acoustic mode in the direction perpendicular to the layers (ΔTA). In fact, in the absence of the inter-layer interactions, the energies of the ΔTA branch would be identically zero, and the $\Sigma TA_{\perp}$ branch would have a dispersion curve of quadratic form near the zone center. MoS_2, which is a semiconductor and therefore has no conduction electrons, exhibits a convex dispersion due to its weakly coupled layer structure [31]. The (ξ 00)Σ_3 phonon branch in $NbSe_2$, which has out-of-plane polarisation, is therefore expected to exhibit a weak convex curvature.

From the data of Figs. 7.17 and 7.18 it is not straightforward to verify whether in addition to a convex curvature, a sharp kink is present. While a convex curvature resulting from weak inter-layer forces will have no effect on the momentum dependent phonon linewidth, a kink originating from a Kohn anomaly will lead to a sharp broadening of the phonons. We have therefore studied the momentum dependent phonon linewidth of the (ξ 00)Σ_3 branch. Due to the relatively small sample size (200 mg) the measurements were restricted to small energies and high temperatures. Figure 7.19 shows the raw results. The data are not corrected for resolution due to the lack of the off-symmetry dispersion data. The increase in the linewidth at small wave vectors is entirely due to resolution effects from the curvature and the mosaic spread of the sample. The dotted line represents the trend of the resolution function, which decreases with q (see Sect. 6.4 in Chap. 6). A jump in the raw linewidths above $\xi = 0.11$ rlu, the same wave vector where the dispersion exhibits a kink, is observed. The statistics however is not enough to draw solid conclusions.

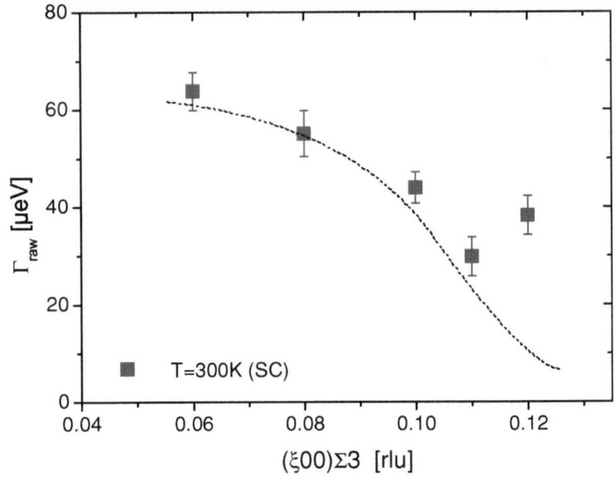

Fig. 7.19 Raw phonon linewidths of the $(\xi 00)\Sigma_3$ branch in $NbSe_2$ measured on the single crystal (SC) at $T = 300$ K. The increase in the linewidth with decreasing ξ is due the curvature and mosaic contributions to the NRSE resolution. The *line* represents the momentum dependency of the instrumental resolution

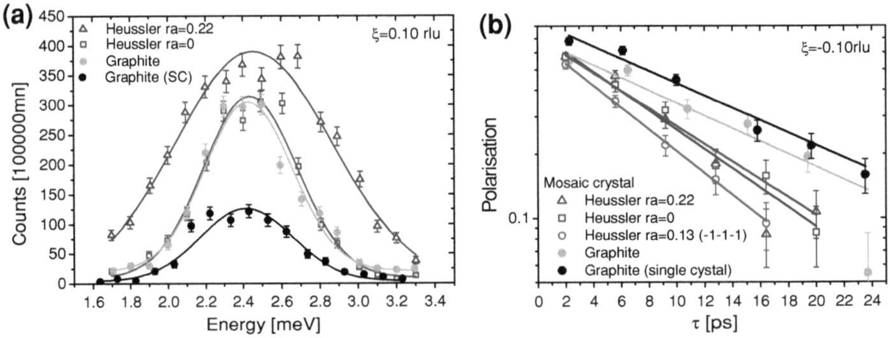

Fig. 7.20 **a** Energy scan of the $\xi = 0.1$ rlu phonon in the single crystal (SC) and the mosaic crystal (MC) at $T = 300$ K. The scans were performed under different spectrometer configuration. ra is the horizontal focusing of the Heussler analyzer. **b** NRSE scans of the same phonon under different spectrometer configurations. $(-1\ -1\ -1)$ represents the scattering sense of the particular scan. All other NRSE scans were performed in the $(-1\ -1\ +1)$ configuration

To perform similar measurements at temperature below and just above T_c, a larger mosaic array of single crystals (MC) was prepared. The sample array was composed of five single crystals, each attached and co aligned on a tiny goniometer (see Fig. 6.6a in Chap. 6). The total weight of the sample array was $\approx$700 mg with an effective mosaic spread of $\eta \approx 10'$. To further enhance the intensity, the graphite analyzer with the supermirror spin polariser was replaced by a horizontally focused Heussler analyzer. Figure 7.20a compares a TAS energy scan of the $\xi = 0.1$ rlu phonon performed with the SC and MC under different spectrometer configurations. A gain of a factor of three in intensity is reached with the mosaic crystal (solid green and black circles). While the intensity is further enhanced by the horizontal focusing (ra $= 0.22$ where ra is the radius of curvature) of the Heussler analyzer, the resulting phonon peak is also broadened.

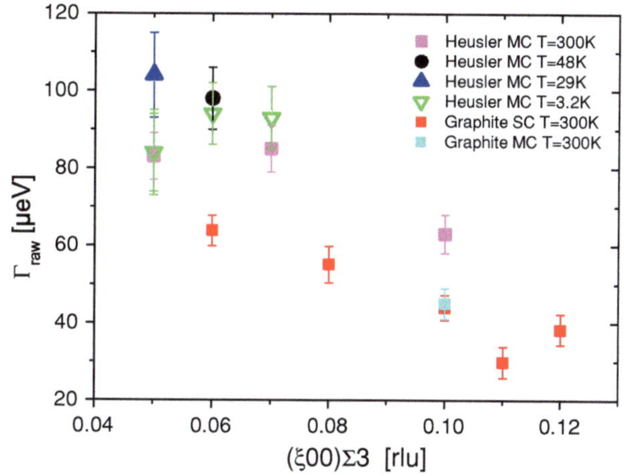

Fig. 7.21 Summary of the raw phonon linewidth of the $(\xi 00)\Sigma_3$ branch in $NbSe_2$ measured under different experimental configurations. The increase in the linewidths with decreasing ξ is mainly a result of the resolution. Measurements under the same experimental condition for the two samples (SC and MC) display the same result ($\xi = 0.10$ rlu)

Figure 7.20b compares the corresponding NRSE scans. The linewidth in the SC and the MC measured with the graphite analyzer (Black and green circles) display similar results ensuing a comparable mosaic spread for the two samples. The slight drop of the initial polarisation for the MC is due to the inhomogeneous distribution of the crystals. The linewidths measured with the Heussler analyzer all show a broadening as a consequence of the worse resolution. Figure 7.21 displays a summary of the momentum dependent linewidths carried out at different temperatures. Even with the large crystal array and the focused analyzer, measurements at low temperature were mainly possible for small ξ. It is important to note that $NbSe_2$ exhibits two critical temperatures: $T_c = 7.2$ K and $T_{CDW} = 33.5$ K. No appreciable temperature dependency across any of the two critical temperatures was observed. This is foreseen since these wave vectors are too small to connect any two points on any of the $2D$ Fermi surface sheets, and thus weakly interact with the electrons. Instead due to the low dimensionality of the Fermi surface, only a few but rather sharp features at nesting wave vectors are expected which require measurements on a wider momentum range. At large ξ (above 0.10 rlu), where we expect to come across a connection between the superconducting gap and the observed kink at room temperature, the intensities were too low for a descent measurement.

A larger crystal array (total weight $\approx 3gr$) is therefore essential for a conclusive experiment.

7.2 High Temperature Superconductor LSCO

Based on the successful observation of e–p linewidths with μeV resolution in conventional superconductors, we extended the research to the more sophisticated high temperature superconductors, whose understanding is still in its infancy.

Unlike the simple elemental superconductors, which exhibit mainly distinct acoustic branches, $La_{2-x}Sr_xCuO_4$ has seven atoms per unit cell and therefore displays 18 optical and 3 acoustic phonon branches extended to energies up to $\approx$80 meV. The focus of the present work was mainly on the low energy acoustic phonons. These branches, however, are highly contaminated by the presence of several low-lying optical modes, which cross and anti-cross the acoustic branches. To avoid spurious effects in the NRSE measurements, it was essential to have a map of the low lying phonon energies prior to the NRSE measurements.

We have investigated the temperature and momentum dependence of the phonon energies and linewidths in $La_{2-x}Sr_xCuO_4$ (LSCO) in the underdoped ($x = 0.08$), optimally doped ($x = 0.14$), and overdoped ($x = 0.22$) samples [32]. Most of the measurements were carried out on the transverse acoustic $(\xi 00)\Delta_4$ branch. Figure 7.22 displays the dispersion of the transverse acoustic $(\xi 00)\Delta_4$ phonon branch for the different dopings measured at $T = 10$ K. The measurements were performed with a fixed $k_f = 3.2\text{Å}^{-1}$ near the (1 0 11) Bragg peak which has the highest structure factor for this phonon branch (see Fig. 7.22b). The low lying optical branches as well as the longitudinal acoustic branch for the undoped sample, taken from [33], are also plotted. To distinguish among the different branches and their eigenvectors, high resolution TAS measurements were carried out on PUMA spectrometer at the FRM-2, which is optimized to measure phonon dispersions rather than linewidths. The measurements were performed with a fixed $k_f = 2.662\text{Å}^{-1}$ near the (0 0 6) Bragg peak to optimize the momentum resolution. Graphite filters were used to filter out the higher harmonics.

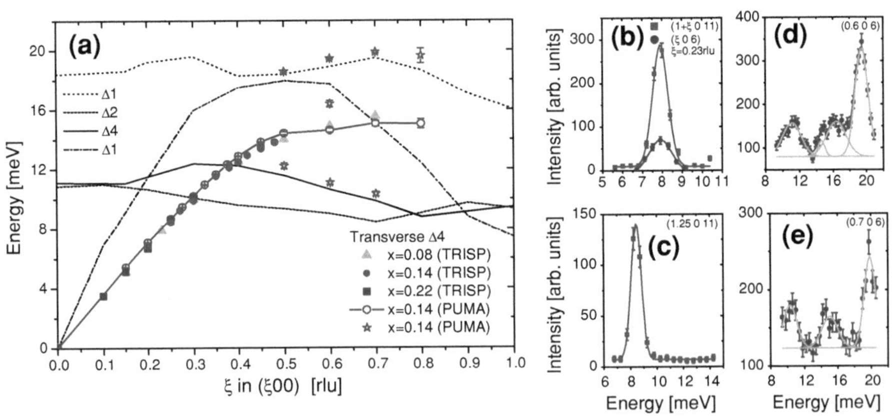

Fig. 7.22 **a** Phonon dispersion of the transverse acoustic $(\xi 00)\Delta_4$ branch for the different samples used in the experiment. The measurements were performed on TRISP and on PUMA spectrometers at the FRM-2. The low lying optical and the longitudinal acoustic branches taken from [33] are represented by the *black lines*. **b** TAS energy scans of the $\xi = 0.23$ rlu phonon at $T = 10$ K measured on TRISP near the (0 0 6) and (1 0 11) Bragg peaks. The difference in intensity is due to the phonon structure factor. **c–e** TAS energy scans near the (1 0 11) carried out on TRISP and near the (0 0 6) carried out on PUMA. The *lines* are the results of fits to Gaussians

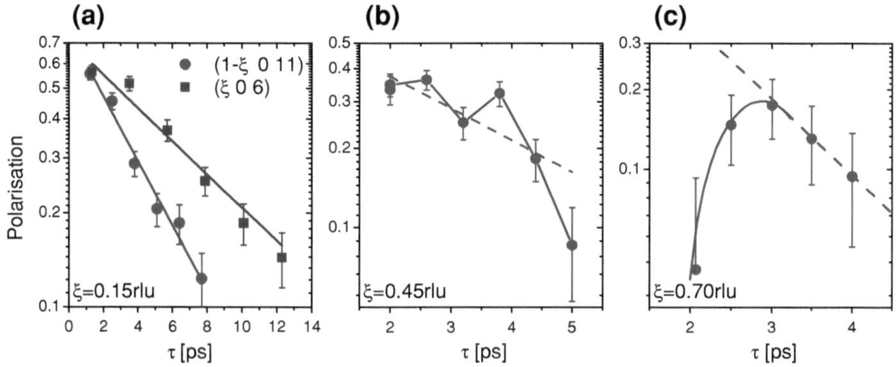

Fig. 7.23 Raw NRSE profile for several selected transverse acoustic $(\xi 00)\Delta_4$ phonons in optimally doped LSCO. The $\xi = 0.15$ rlu phonon displays a single exponential decay (**a**). The difference in the raw linewidths is due to the $q-$ resolution. For $\xi > 0.40$ rlu the NRSE signal is contaminatedby the nearby optical branches (**b**, **c**). The *lines* in (**b**, **c**) are guide to the eye

Figure 7.22d and e displays two sets of energy scans showing multiple phonon groups from the different branches. The extracted energies are plotted in Fig. 7.22a. Except for the Δ_2, all other branches show a transverse component and are visible near the (0 0 6) Bragg peak. Due to their same symmetry, the optical Δ_4 branch anti-crosses the transverse acoustic Δ_4 branch near $\xi \approx 0.4$ rlu leaving a gap in the phonon spectrum.

Due to the mixing of the different branches and the contamination of the acoustic $(\xi 00)\Delta_4$ phonons, NRSE scans showed deviations from a single Lorentzian for $\xi > 0.40$ rlu. Figure 7.23 displays raw NRSE profiles for several $(\xi 00)\Delta_4$ phonons. While the $\xi = 0.15$ rlu phonon shows a single exponential decay (panel a), an appreciable oscillation of the beam polarisation as a function of the spin echo time τ is observed for the $\xi = 0.45$ and 0.70 rlu phonons (panels b, c) which strongly deviate from a single exponential decay. These deviations indicate finite scattering within the TAS ellipsoid from the nearby modes. To avoid spurious effects, the spin echo measurements were confined to low-ξ region of the dispersion. Figure 7.23a also shows the remarkable enhancement of resolution for measurements near the (0 0 6) Bragg peak compared to that of the (1 0 11), however for a huge expense of intensity as shown in Fig. 7.22b.

The temperature dependence of the energy and the linewidth of two selected phonons in optimally doped LSCO is shown in Fig. 7.24. The measurements were performed near the (0 0 6) Bragg peak with fixed $k_f = 2.6\text{Å}^{-1}$. The energies and the linewidths are extracted from the relative phase and polarisation shifts of the spin echo signal (Fig. 7.24e). Both quantities follow the trend of the structural phase transition in optimally doped LSCO from a HTT to LTO phase (Fig. 7.24f). While the $\xi = 0.20$ rlu phonon does not show a remarkable anomaly across $T_c \approx 35$ K, the $\xi = 0.15$ rlu phonon exhibits a superconductivity induced broadening of $\approx 20\mu\text{eV}$ below T_c (Fig. 7.24d). Figure 7.25 summarizes the temperature

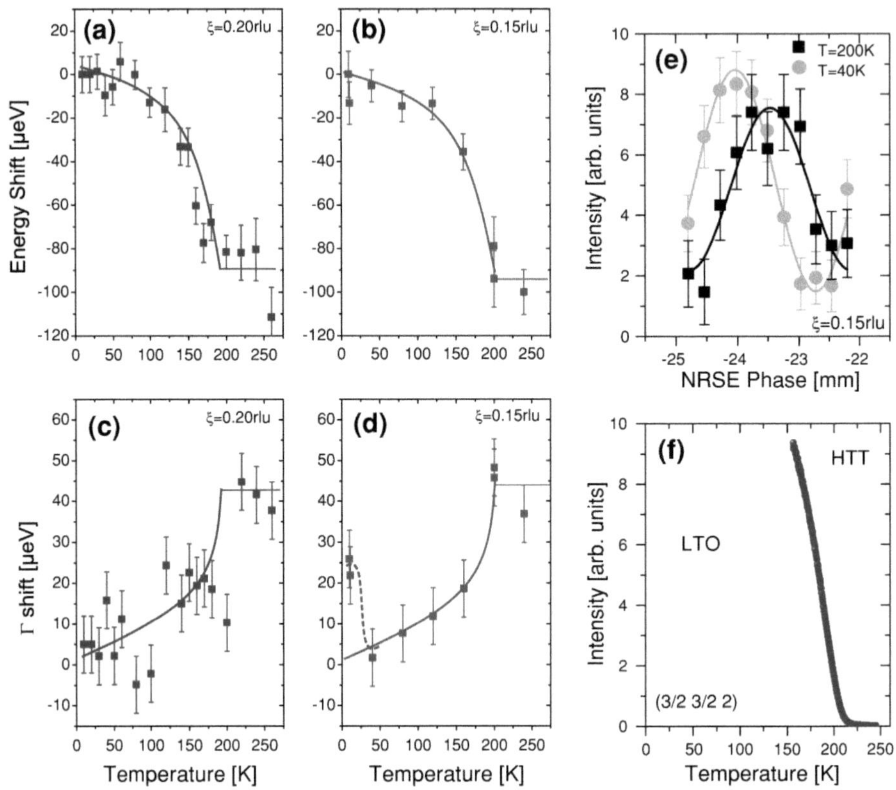

Fig. 7.24 Energy (**a**, **b**) and linewidth (**c**, **d**) shifts of two selected transverse acoustic $(\xi 00)\Delta_4$ phonons in optimally doped LSCO. The *lines* are guide to the eye. **e** Spin echo intensity modulation measured at $\tau = 10.1$ ps at 40 and 200 K. The phase shift of the signal is proportional to the energy shift, whereas the change in the polarisation depends on the linewidth shift. **f** The (3/2 3/2 2) supperlattice peak in optimally doped LSCO which is allowed in the low temperature orthorhombic (LTO) phase and forbidden in the high temperature tetragonal (HTT) phase

and momentum dependent phonon linewidth in optimally doped LSCO. The NRSE profiles, within this momentum range, are well described by single exponentials (panel b). The measurements were performed near the (0 0 6) Bragg peak to optimize the resolution. Within the momentum range covered in the experiment, linewidth broadening of ≈ 40 μeV is observed between 50 and 280 K ensuing from the structural phase transition at $T_o \approx 210$ K. Both spectra however also exhibit an additional momentum dependent broadening for $\xi > 0.20$ rlu. This linewidth broadening can be due to *e–p* interaction, which weakens for $\xi < 0.15$ rlu presumably due to the geometry of the Fermi surface (where phonons with wave vectors $\xi < 0.15$ rlu do not have enough momentum to scatter electrons across the antinodal part of the Fermi surface). At $T = 10$ K in the superconducting state, the $\xi = 0.25$ rlu phonon exhibits a slight linewidth narrowing. This is expected since this phonon lies below the superconducting energy gap $2\Delta_q$

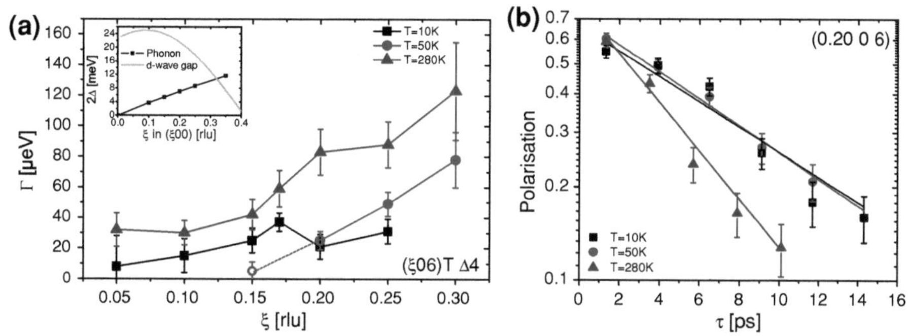

Fig. 7.25 **a** Momentum dependence of the intrinsic phonon linewidth in optimally doped LSCO at selected temperatures. The *lines* are guide to the eye. The *open red circle* represents the linewidth extracted from Fig. 7.24. The *inset* shows the d-wave superconducting energy gap which crosses the phonon dispersion at $\xi \approx 0.32$ rlu. **b** Raw NRSE profile of the $\xi = 0.20$ rlu phonon at selected temperatures. The *lines* are fit to exponentials

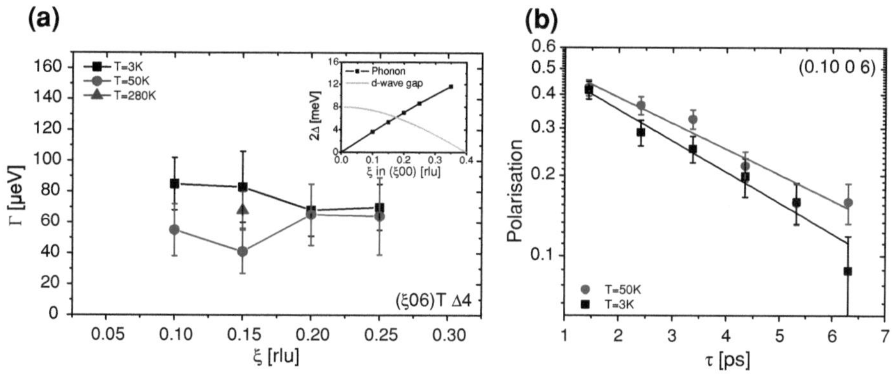

Fig. 7.26 **a** Momentum dependence of the intrinsic phonon linewidth in overdoped LSCO at selected temperatures. The *lines* are guide to the eye. The *inset* shows the d-wave superconducting energy gap which crosses the phonon dispersion at $\xi \approx 0.17$ rlu. **b** Raw NRSE profile of the $\xi = 0.10$ rlu phonon at selected temperatures. The *lines* are fit to exponentials

(inset of Fig. 7.25a), where the *e–p* interaction is suppressed due to the stability of the Cooper pairs. Yet the $\xi = 0.15$ rlu phonon, which is expected to lie far below $2\Delta_q$, shows an anomalous broadening below T_c. Similar anomalous linewidth broadening in the superconducting state was observed in the overdoped and the underdoped samples for $\xi < 0.20$ rlu (Figs. 7.26, 7.27). This broadening, which seems to be induced by superconductivity, can not be understood within the excepted gap magnitudes for LSCO [34, 35]. Such a broadening would be expected only for phonons which lie above $2\Delta_q$. While the optimally doped and underdoped samples show linewidth narrowing below 0.20 rlu, the overdoped sample exhibits linewidth broadening even for small phonon wave vectors. This can be understood from the geometry of the Fermi surface, where in the overdoped

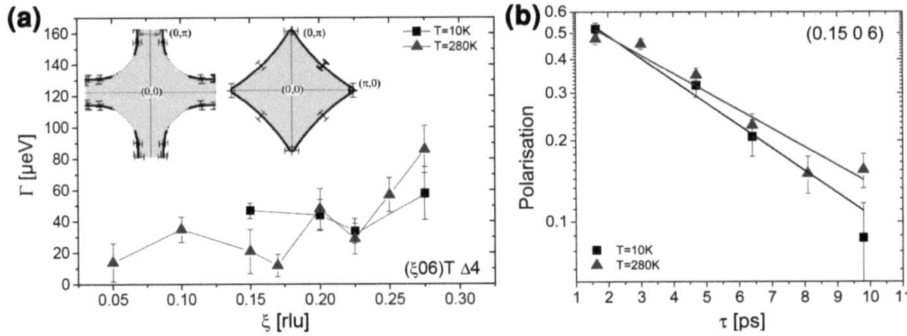

Fig. 7.27 **a** Momentum dependence of the intrinsic phonon linewidth in overdoped LSCO at selected temperatures. The *lines* are guide to the eye. The *inset* shows the Fermi surface in underdoped (*left*) and overdoped (*right*) LSCO [34]. In the underdoped case, phonons with wave vector smaller than the red arrow ($\sim$0.20 rlu) cannot scatter electrons along the 100 direction. This restriction does not hold for the overdoped case. **b** Raw NRSE profile of the $\xi = 0.15$ rlu phonon at selected temperatures. The *lines* are fit to exponentials

regime the antinodal (π, 0) region becomes connected (see inset of Fig. 7.27a). The momentum and the doping dependence of the linewidth at low temperatures supports the *e–p* nature as the origin to the observed broadening. Yet the anomalous temperature dependency across T_c remains unclear.

In addition to the $(\xi 00)\Delta_4$ branch, measurements of the transverse acoustic $(\xi 00)\Delta_3$ and $(\xi\xi 0)\Sigma_4$ branches were also performed. Figure 7.28 shows the dispersion of the transverse acoustic $(\xi 00)\Delta_3$ branch in optimally doped LSCO. The measurements were performed with a fixed $k_f = 2.5\text{Å}^{-1}$ near the (2 0 0) Bragg peak at $T = 10$, 50 and 250 K. A considerable hardening is observed at 250 K, presumably due to the structural transition from LTO to HTT at $T_o \approx 210$ K (Fig. 7.24f). The low lying Δ_2 and Δ_4 symmetry optical branches taken from [33] are also plotted. Due to their different symmetry, both branches cross the Δ_3 branch. Scattering from these optical branches was not observed in the measurements near the (2 0 0) Bragg peak, as is obvious by the single Gaussian peaks in Fig. 7.28a and b. NRSE scans for two $(\xi 00)\Delta_3$ phonons are shown in Fig. 7.29. The $\xi = 0.45$ rlu phonon, shows an oscillation of the polarisation as a function of the spin echo time τ, apparently due to contamination from the nearby optical Δ_2 branch. Though signatures of the Δ_2 branch were not observed in the TAS energy scans, this can be due to the fact that in a TAS focusing case, a weak scattering from the Δ_2 branch leads to a constant background, which is hardly detectable. On the other hand the $\xi = 0.30$ rlu phonon, which lies away from the optical branches, displays a single exponential behavior (panel a). Nevertheless, superconductivity induced linewidth broadening was not observed across T_c.

Figure 7.30b–d display several TAS energy scans of the $(\xi\xi 0)\Sigma_4$ phonon branch in optimally doped LSCO. The measurements were performed at $T = 10$ K with a fixed $k_f = 2.2\text{Å}^{-1}$ near the (0 0 6) Bragg peak which displays the highest phonon structure factor (panel a). The acoustic $(\xi\xi 0)\Sigma_4$ phonon dispersion is

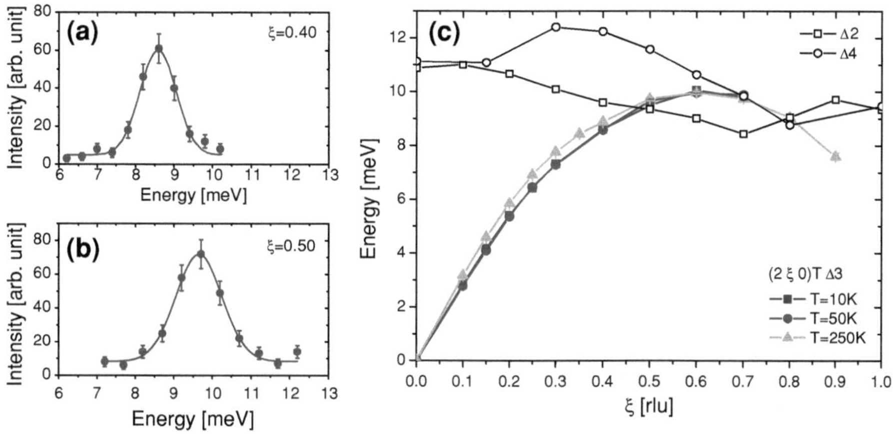

Fig. 7.28 **a**, **b** TAS energy scans of the transverse acoustic $(\xi 00)\Delta_3$ phonons in optimally doped LSCO. The *lines* are fit to Gaussians. **c** The transverse acoustic $(\xi 00)\Delta_3$ phonon dispersion at selected temperatures. The *black lines* are the low lying optical Δ_2 and Δ_4 branches [33] which cross the Δ_3 branch

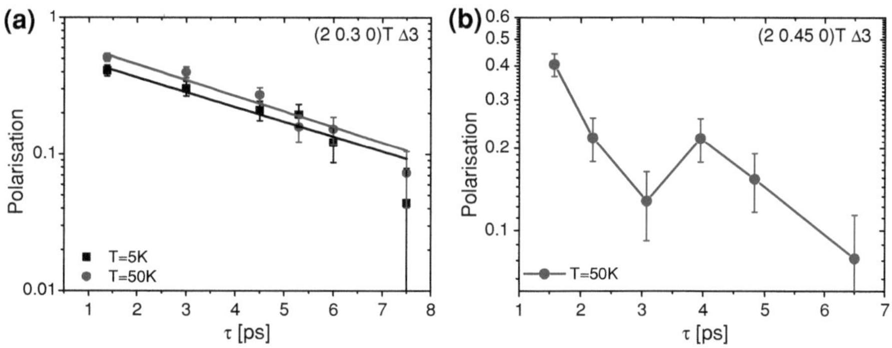

Fig. 7.29 **a** Raw NRSE profile of selected transverse acoustic $(\xi 00)\Delta_3$ phonons in optimally doped LSCO. The *lines* in (**a**) are fits to exponentials, whereas the line in (**b**) is guide to the eye showing a multi Lorentzian feature

shown in panel e. Data for the transverse acoustic Σ_3 branch taken from [36] are also plotted. While the two branches are nearly overlapping, the Σ_3 branch should have a zero structure factor near the (0 0 6) Bragg peak, since it is polarised along the $\varepsilon = (\bar{1}10)$ direction, out of the scattering plane, i.e $(\mathbf{Q}.\varepsilon)^2 = 0$. Indeed the energy scans display single phonon groups corresponding to the Σ_4 branch. Figure 7.31 shows NRSE scans for two $(\xi\xi 0)\Sigma_4$ phonons in optimally doped LSCO ($x = 0.14$). The measurements were performed with fixed $k_f = 2.5\text{Å}^{-1}$. An appreciable oscillation of the beam polarisation as a function of the spin echo time τ is, ones again, observed for the $\xi = 0.05$ rlu phonon (panel a) which strongly deviates from a single exponential decay. The two acoustic modes (Σ_3 transverse

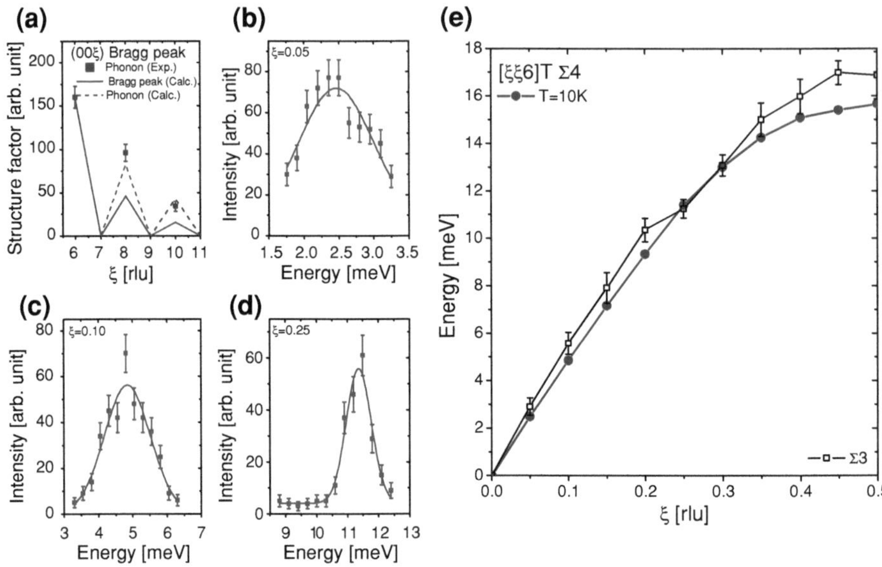

Fig. 7.30 **a** Structure factor of transverse acoustic $(\xi\xi 0)\Sigma_4$ phonons measured near different Bragg peaks. **b–d** Selected $(\xi\xi 0)\Sigma_4$ phonon groups in optimally doped LSCO measured at $T = 10$ K all showing a single Gaussian. **e** Dispersion of the $(\xi\xi 0)\Sigma_4$ branch. The Σ_3 acoustic branch taken from [36] is also plotted

and Σ_1 longitudinal), which might lie within the TAS ellipsoid centered at $\xi = 0.05$ rlu, should in principle have zero structure factors near the (0 0 6) Bragg peak. Nevertheless, due to the finite out-of-plane resolution, scattering from the off-symmetry phonons of those branches seems to be the source of signal contamination. The NRSE profile for the $\xi = 0.10$ rlu phonon (panel b), displays a single exponential decay at both temperatures. Yet the remarkable difference of the initial polarisation P_o for the two temperatures (inset), and the slight higher polarisation around $\tau = 6$ ps compared to the $\tau = 4$ ps at $T = 5$ K are indications of a weak but a finite contamination from a nearby phonon group. Even though the intrinsic linewidths do not represent single Lorentzians, both phonons exhibit a significant temperature dependence between 5 and 50 K which can be attributed to superconductivity. The phonon linewidth across T_c of this particular phonon branch was investigated by Chou et al. [37] in optimally doped LSCO. With the limited energy resolution of the TAS, they were able to establish an upper limit of 100 μeV on the change in the linewidth across T_c. From the present data we can estimate a $\Gamma^{e-p} \approx 40$ μeV as an upper limit.

It is worth mentioning that LSCO develops a domain structure, known as 'twins', in the orthorhombic phase. Such a domain structure which leads to the splitting of the Bragg peaks and therefore the acoustic phonon dispersion, also leads to oscillation of the NRSE signal. Nevertheless, the magnitude of this splitting (≈ 0.005 rlu corresponding to a maximum splitting of ≈ 20 μeV of the acoustic dispersion) will lead to an oscillation of period $\Delta\tau > 200$ ps. Within the

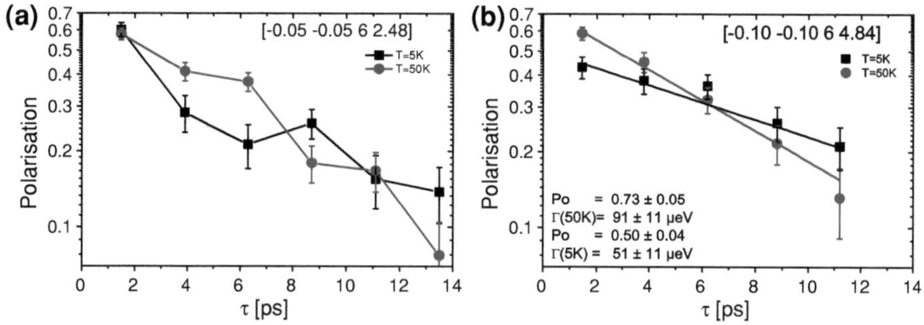

Fig. 7.31 Raw NRSE profile of the transverse acoustic $(\xi\xi 0)\Sigma_4$ phonons with $\xi = 0.05$ rlu (**a**) and $\xi = 0.10$ ru (**b**). Both phonons show contamination from nearby phonons. The inset of (**b**) gives the initial polarisation (P_o) and the raw linewidths (Γ) extracted from exponential decay fits

experimentally covered τ range (0–15 ps), the polarisation can be assumed to be constant.

In summary, we have observed signatures of *e–p* interaction in underdoped, optimally doped, and overdoped LSCO. Momentum dependent broadening $\approx$40 μeV were observed for the transverse acoustic $(\xi 00)\Delta_4$ branch. This branch however displayed an anomalous temperature dependency across T_c. The lower transverse acoustic Δ_3 branch along the same direction did not exhibit any anomaly across T_c. Clear signature of superconductivity induced broadening across T_c was only observed for the transverse acoustic $(\xi\xi 0)\Sigma_4$ phonons.

References

1. S. Shapiro et al., Phys. Rev. B **12**, 4899 (1975)
2. J.D. Axe et al., Phys. Rev. B **8**, 1965 (1973)
3. J.D. Axe et al., Phys. Rev. Lett. **30**, 214 (1972)
4. Y. Nagakawa et al., Phys. Rev. Lett. **11**, 271 (1963)
5. R.C. Dynes et al., Phys. Rev. Lett., **41**, 1509 (1978)
6. The calculations were carried out by Lilia Boeri from O.K. Anderson department at the Max Planck Institute for Solid State Research
7. W. Wolf et al., Science-Poland **23**, 365 (2005)
8. D.J. Scalapino, Science **319**, 1492 (2008)
9. A. Furrer et al., Phys. Status Solidi **42**, 821 (1970)
10. B.N. Brockhouse et al., Phys. Rev. **128**, 1099 (1962)
11. R. Stedman et al., Phys. Rev. **162**, 161 (1967)
12. R. Youngbloud et al., Solid State Commun. **27**, 1433 (1978)
13. D.L. Waldorf, Bull. Am. Phys. Soc. **5**, 170 (1960)
14. A.W. Overhauser, Phys. Rev. **128**, 1437 (1962)
15. X.M. Chen et al., Phys. Rev. B **39**, 10570 (1989)
16. A.W. Overhauser et al., Phys. Rev. B **47**, 14338 (1993)
17. A. Dal Corso, J. Phys. Condens. Matter **20**, 445202 (2008)
18. R.F. Gasparovic et al., Solid State Commun. **4**, 59 (1966)
19. J.R. Anderson et al., Phys. Rev. **139**, A1459 (1963)

20. H.J. Maris et al., Phys. Rev. Lett. **25**, 220 (1970)
21. L.P. Pitayevski et al., Phys. Rev. B **14**, 263 (1976)
22. The samples were grown by the floating zone method in C.T. Lin's Department at the Max Planck Institute for Solid State Research
23. J.P. Carbotte, Rev. Mod. Phys. **62**, 1029 (1990)
24. M. Hansen, *Constitution of Binary Alloys* (McGrawHill, New York, 1958)
25. S.V. Borisenko et al., Phys. Rev. Lett. **100**, 196402 (2008)
26. G. Gruner, *Density Waves in Solids* (Addison-Wesley, Massachusetts, 1994)
27. T. Yokoya et al., Science **294**, 2518 (2001)
28. T. Kiss et al., Nat. Phys. **3**, 720 (2008)
29. D.E. Moncton et al., Phys. Rev. B **16**, 801 (1977)
30. The samples were grown by Helmut Berger in Lausanne Switzerland
31. N. Wakabayashi et al., Phys. Rev. B **12**, 659 (1975)
32. The crystals had a mass of 10–12 g and a mosaicity of typically 5′. They were kindly provided by K. Yamada, Tohoku University Tokyo
33. L. Pintschovius, Phys. Status Solidi B **242**, 30 (2005)
34. A. Ino et al., Phys. Rev. B **65**, 094504 (2002)
35. K. Terashima et al., Phys. Rev. Lett. **99**, 017003 (2007)
36. P. Böni et al., Phys. Rev. B **38**, 185 (1988)
37. H. Chou et al., Phys. Rev. B **42**, 4272 (1990)

Chapter 8
Conclusion

Neutron resonance spin-echo spectroscopy was used to investigate the lifetimes of acoustic phonons in conventional and unconventional superconductors. A major improvement of the energy resolution by almost two orders of magnitude, compared to conventional triple-axis spectroscopy, was achieved, which made this research possible.

In elemental superconductors Nb and Pb, the electron–phonon contribution to the phonon lifetimes were resolved with μeV resolution along the high symmetry directions. The results confirmed the ab-initio lattice-dynamical calculations [1, 2]. The superconducting energy gap $2\Delta(T)$ was accurately determined form the temperature and momentum dependent discontinuities in the phonon lifetimes. The temperature dependence of the extracted gap values agree well with the BCS expression of the superconducting gap [3]. Surprisingly, in both elements, the low-temperature gap converged with sharp Kohn anomalies originating from Fermi-surface nesting. While the gap value, $2\Delta(0)$, along the high-symmetry directions probed experimentally, differed by 8 and 20% in Nb and Pb, respectively, this gap anisotropy mirrored the normal state Kohn anomalies in momentum space. The coincident momentum dependence of both quantities suggests that the superconducting gap and its momentum dependent anisotropy at low temperatures are controlled by the locus of Kohn anomalies in the acoustic phonon spectrum. Based on ab-initio density-functional calculations it appears plausible that this coincidence extends over the entire Brillouin zone. This mechanism of gap anisotropy thus provides a new perspective to this long-standing problem. While the same ingredients that had been discussed in prior theoretical work are involved, namely anisotropies of the Fermi surface and the phonon spectrum, they are combined in a surprisingly simple manner.

Further experiments, carried out on two Pb–Bi samples with 3 and 10% Bi concentration, revealed the expansion of the Fermi surface and the consecutive displacement of the Kohn anomalies to larger wave vectors. In both alloys, the superconducting energy gap was found to follow and lock to the Kohn anomaly

P. Aynajian, *Electron–Phonon Interaction in Conventional and Unconventional Superconductors,* Springer Theses, DOI: 10.1007/978-3-642-14968-9_8,

enhancing both the gap magnitude and the T_c. These experiments on pure elements and alloys demonstrate that the low temperature limit of the superconducting energy gap locks in to the lowest Kohn anomaly in transverse acoustic phonons. A locking mechanism was proposed by Scalapino [4] which includes feedback effects of superconductivity on the phonon spectrum, but further theoretical work is required to provide a definitive, quantitative explanation of the "lock-in" effect. Lastly, work on other elemental and compound superconductors is required to assess the generality of the lock-in mechanism. While the effect is clearly inoperative for superconductors with very low transition temperatures, whose gap is far below the lowest-energy Kohn anomaly, its presence in other superconductors with known or suspected gap anisotropy, such as elemental tin [5] or gallium [6], is an interesting subject of further investigation.

Based on the successful observation of *e–p* linewidths with μeV resolution in conventional superconductors, the research was extended to the high temperature superconductor $La_{2-x}Sr_xCuO_4$ (LSCO). The temperature and momentum dependence of the transverse acoustic phonon energies and linewidths in underdoped ($x = 0.08$), optimally doped ($x = 0.14$), and overdoped ($x = 0.22$) LSCO samples were investigated. Due to the contamination of the acoustic branches, by the presence of several low-lying optical modes, measurements were confined to specific regions of the dispersion. Signatures of superconductivity induced broadening of $\Gamma_{e-p} \approx 40$ μeV above T_c was observed for the transverse acoustic $(\xi\xi0)\Sigma_4$ phonons. The lowest transverse acoustic $(\xi00)\Delta_3$ branch did not exhibit any anomaly (within the error bars of $\approx 10\,\mu$eV) across T_c, whereas the higher acoustic Δ_4 branch along the same direction displayed an anomalous temperature dependent broadening below T_c for all measured samples. The observed temperature dependence suggests that the gap magnitude which these phonons probe is quite small. Momentum and doping dependent linewidth broadening of $\approx 40\,\mu$eV were also observed for this same branch. In addition, the acoustic $(\xi00)\Delta_4$ phonons displayed softening of $\approx 100\,\mu$eV and broadening of $\approx 50\,\mu$eV at the HTT to LTO structural phase transition, whereas the acoustic $(\xi00)\Delta_3$ phonons displayed hardening of $\approx 400\,\mu$eV.

An upper limit of $\Gamma_{e-p} \approx 50\,\mu$eV for acoustic phonons can be concluded for all the dopings, which is comparable to the *e–p* contribution to the linewidths observed in conventional superconductors with much smaller transition temperatures. Clearly, therefore coupling of electrons to the acoustic phonons alone cannot explain the high transition temperatures in cuprates. Contributions of the optical phonon modes to the pairing interaction are an important subject for further investigations.

References

1. The calculations were carried out by Lilia Boeri from O. K. Anderson Department at the Max Planck Institute for Solid State Research.
2. A. Dal Corso, J. Phys. Condens. Matter. **20**, 445202 (2008)

3. J. Bardeen et al., Phys. Rev. **108**, 1175 (1957)
4. D.J. Scalapino, Science **319**, 1492 (2008)
5. P.L. Richards, Phys. Rev. Lett. **7**, 412 (1961)
6. K. Yoshihiro et al., J. Phys. Soc. Jpn. **28**, 262 (1970)

Vita

Pegor Aynajian
Department of Physics, Princeton University
40 Jadwin Hall, Princeton, NJ 08544, USA
Phone: +1-609-751-1278
Fax: +1-609-258-1006
e-mail: aynajian@princeton.edu

Born on the 28th of October 1981 in Beirut, Lebanon. Bachelor (2002) and *Maitrise* (2003) degree in physics from Lebanese University, Lebanon. Masters in physics (2005) from Stuttgart University, Germany. Ph.D. in experimental condensed matter physics (2009) from Stuttgart University and the Max Planck Institute for Solid State Research, Germany. Post doctoral fellowship (Materials Research Science and Engineering Centers—MRSEC) (since 2009) at the Joseph Henry Laboratories of Physics, Princeton University, USA.

Selected Publications

"Visualizing the Formation of the Kondo Lattice and the Hidden Order in URu_2Si_2"P. Aynajian, E. da Silva Neto, C. V. Parker, Y. Huang, A. Pasupathy, J. Mydosh, A.Yazdani, *Proc. Nat'l. Acad. Sci. (PNAS)*, (2010).

"Energy Gaps and Kohn Anomalies in Elemental Superconductors" P. Aynajian,T. Keller, L. Boeri, S. M. Shapiro, K. Habicht, B. Keimer, *Science* 319 (2008), 1509.

"The Triple Axis Spin-Echo Spectrometer TRISP at the FRMII" T. Keller, P. Aynajian, S. Bayrakci, K. Buchner, K. Habicht, H. Klann, M. Ohl, B. Keimer, *Neutron News*, Volume 18, Issue 2 April 2007, pages 16–18.

"Momentum-Resolved Electron–Phonon Interaction in Lead Determined by Neutron Resonance Spin-Echo Spectroscopy" T. Keller, P. Aynajian, K. Habicht, L. Boeri, S.K. Bose, and B. Keimer, *Phys. Rev. Lett.* 96 (2006), 225501.

MIX
Papier aus verantwortungsvollen Quellen
Paper from responsible sources
FSC® C105338

If you have any concerns about our products,
you can contact us on
ProductSafety@springernature.com

In case Publisher is established outside the EU,
the EU authorized representative is:
Springer Nature Customer Service Center GmbH
Europaplatz 3, 69115 Heidelberg, Germany

Printed by Libri Plureos GmbH
in Hamburg, Germany